Kishori Aird

ESSENCE

EACH OF US HAS A UNIQUE AND ESSENTIAL FREQUENCY .
ARE YOU AWARE OF YOUR OWN FREQUENCY?

EDITOR: Copyright – © Institut Kishori, Inc.
P.O. Box 252
Magog (Quebec) JIX 3W8
CANADA
Tel.: 819-868-1284, Fax: 819-868-9007
Web site: www.Kishori.org
Email: kishori@kishori.org

Editing: Louise Drolet
Translation: Rose-Anne Chabot and Françoise McNeil
Proofing: Anna Birtles
Layout and illustration: Devin Young

First printing: May 2006
ISBN 2-9807441-6-6
ISBN 978-2-9807441-6-7
Legal deposit – Bibliothèque nationale du Québec, 2006
Legal deposit – National Library of Canada, 2006

DISTRIBUTION
CANADA
PHOENIX DISTRIBUTORS, SURREY, B.C.
1 800 563-6050, WWW.PHOENIXDISTRIBUTORS.BIZ

UNITED STATES
NEW LEAF DISTRIBUTING COMPANY, LITHIA SPRINGS, GA 301227-1557
1 800 326-2665, WWW.NEWLEAF-DIST.COM

OTHER
BRUNO SIMARD
819-868-0595, BRUNO@KISHORI.ORG

PRINTED IN CANADA

Table of Contents

PART III. ESSENCE

PART IV. THE ASCENSION PROCESS

FOREWORD

Essence is the third book that completes a series that also includes *DNA Demystified* and *DNA and the Quantum Choice*. There is no need to have read the previous two before starting this one.

While reviewing the genetic reprogramming protocols introduced in the first two volumes, I noticed a metaphysical philosophy and direction emerging from these texts. In this book, I will review the concepts defined in the previous volumes and elaborate on themes inherent to the quantum way of thinking, the notion of zero point and the power of conscious intentions. Initially, this book was not meant to be an instruction manual, like the previous volumes. Nonetheless, I let myself get carried away and, starting in Part II, you will find exercises with lists of questions. I invite you to read them, think about them and answer them in your own way. At the beginning of each exercise, I have suggested various ways to proceed: *After answering the following questions and establishing your individual profile, you can choose among various tools such as prayer, meditation or DNA reprogramming to put the situation back at zero point.*

I could have put these questions into text form and skipped the instructional steps, but the teacher in me surfaced and would not be silenced. Do not hesitate to use these exercises as you see fit. For some people, reading the protocols that are included in the first two volumes or the exercises found in this book automatically triggers a process that is independent of their rational thinking. Others, however, take pleasure in following the step-by-step instructions. In the quantum universe, a universe of coexistence, all roads lead to Rome.

INTRODUCTION

In *DNA Demystified*, I developed a new thinking process that allows us to understand how we can reprogram our helixes through conscious intentions stated at zero point. It was important, first of all, to define the power of intention and the fact that this power draws on both the positive and negative poles of our mental process. When we state an intention that includes both a positive and a negative pole, the charge of this intention is balanced and the magnetic power of our command is increased. This approach takes place outside of any duality and includes both positive and negative emotions. Our intentions are thus guaranteed to achieve maximum effectiveness. We attain zero point by embracing both darkness and light. This is where we need to start to modify the structure of our DNA using the power of intention at zero point.

Zero point is positioned in a space that exists beyond positive and negative polarities. In this space, holy and evil, light and darkness, good and bad, as well as all other dualities, coexist in equilibrium. Zero point is neither neutral nor static. It is multidimensional and in constant movement. It is the center of a space that is in constant change because the positive force of light and the negative force of darkness coexist without canceling each other, despite their opposing polarities.

The new paradigm I wish to outline in this book is based on the concept of zero point and on the integration of the negative magnetic charge of our emotions. Integrating negative charges is a powerful process. Imagine the intensity of an anger-based charge that has been accumulating for years against a violent parent. If we could measure the energetic intensity of this anger, we would obtain an impressive magnetic charge. Once this charge has been integrated into our DNA, it becomes a strong negative pole to which we can match a corresponding positive pole without having to feel the initial anger. From here on, this magnetic concept will constitute our power to manifest our intentions.

When we apply the concept of zero point, we use the magnetic charge of a negative emotion to represent one pole of a magnet. Since it coexists with a positive charge, it generates a magnetic field. To physically manifest our desires on Earth, we have to associate our commands with a magnetic field. For example, if we fear being the victims of a terrorist attack while on a trip, we can also command, by the power invested in us by this fear, that our trip occur in complete safety. Issuing a positive command that is balanced by a negative charge results in a command that has greater impact.

In *DNA and the Quantum Choice*, I pursued the zero point theory, and looked into the coexistence of quantum realities. This theory forces us to transform our mental process from one that is based on the **two helixes** binary model, to one that is based on a model that is circular, with **thirteen helixes** at zero point. I also developed protocols related to transmutation, ascension, multidimensionality and self-love.

The goal of this new book is to have readers explore a new paradigm outside of duality and create a state of being that is positioned beyond the limits imposed by their past experiences. Unlike books on self-growth and self-esteem, we will adopt a circular way of thinking that is not affected by our default programs and we will explore the art of maintaining a vibrational rate that is in harmony with the frequency rate of our individual essence.

To change the model by which we used to judge reality, we will obviously have to abandon our old model of achievement and self-improvement. We will explore self-acceptance at zero point. My goal is to illustrate how we can make use of all aspects of ourselves and have them coexist. We no longer need to get rid of our outdated beliefs or seek perfection to gain balance.

If we learn to make new choices by making the most of the energy of our strengths and weaknesses, we will put an end to the feeling of powerlessness that cripples our society at the moment. This new paradigm possibly represents our last chance to change the course of history! It is through this new paradigm that we will attain a new stage of evolution in which we will stop aiming for social or professional success. Instead, we will strive to maintain the frequency rate of our individuality.

By opening ourselves up to the bold new ideas in the following chapters, we will begin to understand the new scientific, emotional and spiritual philosophy of zero point! All that matters at the moment is that we become fully ourselves and that we allow all our aspects to vibrate at the unique frequency rate of our essence.

Although **essence** is a word laden with meaning, we would be hard pressed to define it. We know that essence exists because we see it leave the body at the moment of death. We know that there is a vital energy that keeps us alive, but we are not in relation with it. We might think that our essence corresponds to our personality or to our way of being. But in fact, our essence is a vibrational frequency, endowed with a tonality, a rhythmic undulation and a light wave that carries our personal signature. It is not simply a state of being; it is a vibrational frequency that pulsates inside us. The more I experience this wavelength, the more I define myself as the guardian of this frequency. Instead of introducing myself in social settings by giving my educational background, the names of my children and a list of my achievements, I want to be able to introduce myself as a person who is striving to maintain the frequency rate of her essence.

By identifying with a frequency rate specific to my essence, self-love will automatically increase. The combination of these two experiences is what opens the way to ascension.

PART I
THE UNFOLDING OF A LIFE PLAN

Regardless of any particular project we are involved in, we have to take our needs into account because, if we deny them, they will be sure to have a determining influence on our lives. I am increasingly convinced that life follows stages that are more or less the same for all humans, regardless of our age. Of course, life's circumstances are different for each and everyone. At 28, someone may be married and raising young children, or pursuing a flourishing career, whereas, twenty years later, that person may be in the middle of changing careers or starting a new life in a new place. Regardless of the scenario, the underlying pattern remains the same.

We are governed by **default programs**–or unconscious programs. They have to be put at zero point at all costs. Since we have a tendency to make projections, it is important to establish healthy **boundaries**. If we concern ourselves with the future shape of our life plan without clarifying our initial **intentions**, there is a good chance that this will result in failure. To achieve our objectives, we will have to **support** ourselves during moments of doubt, **stop self-criticism** and avoid the trap of hyper-**achievement**. Yet this is not possible unless we discover who we are.

My experience has shown me that our needs and our preferences provide us with valuable clues in finding out who we are. This is why we will adopt a "quantum" way of thinking in which our weaknesses coexist with our strengths, and our bright sides with our dark sides. We will adopt a way of thinking that takes our individual needs into account. Our lives will then be propelled with dynamic movement instead of continuously reproducing unproductive scenarios.

CHAPTER 1
THE NEW DNA MODEL

INFLUENCE OF THE DNA'S ENVIRONMENT

Humanity is at a turning point between two eras: on the one hand, we despair of the violence and ecological tragedies spreading across our planet; on the other hand, biology and physics are providing endless innovations and new models that stimulate our imagination. Isn't it interesting to see how easily old structures can be altered during periods of turbulence? In such moments, our collective thinking becomes receptive to unusual concepts.

We live in an era when even **the truth** could be redefined. Every branch of science is redefining its parameters, especially quantum physics, which states that matter is no longer the solid entity we thought it was, and that concrete reality is determined by our thoughts.

After the superstring theory, which redefines the nature of matter based on its frequency rate, genetics is now poised to modify its basic paradigm. Indeed, 50 years after it was discovered, DNA now appears to be a network of sequential dynamic reactions, sustained and influenced by its environment.

Published in the December 2004 issue of *Science et Vie*, the article "Tracking DNA's Secret Code," challenges all our assumptions on the subject. The author declares that "the theoretical framework that has allowed us to account for life up to now, no longer holds."[1] In the standard model, each DNA molecule corresponds to a specific RNA molecule that produces a specific protein. Now, scientists have discovered that the current model rather resembles that of quantum physics. A single RNA molecule can produce different proteins. Those parts of random DNA that do not have genes or are not coded, and which we also call **junk DNA**, play a role in this equation. But what is most interesting is that genes can influence one another and function as a network. In this new biological model, the source of the genetic response to a signal from the cellular environment is not one-sided. "This experiment demonstrates that RNA and proteins sometimes have a direct effect on the genome. Sometimes even on the gene that produced them."[2] In the old DNA-RNA-protein model,

1 "Sur la piste du code secret de l'ADN". *Science et vie,* Decembre 2004. Translated from the original. p. 52.
2 Ibid., p. 55.

with its specific functions, the elements located at the very end of the chain can also command the DNA.

What governs this process? After fifteen years of research on the human genome, scientists are beginning to understand that our genetic code is not a fixed structure. The dynamics governing the living process are inter-relational. An X gene activated long enough can set off a Y gene, and this activation of XY will lead to the activation of Z. This intermingling of genetic reactions shatters the concept of an established order set up by Watson and Crick in 1953. Matter is not fixed and discoveries in biology continue to corroborate this model. "The deployment of life may well be governed by a principle of disorder, between the well-ordered microstructure of the genes in DNA and the well-ordered macrostructure of living organisms. And it is not based on several thousands of genes but on billions and billions of molecular interactions..."[3]

It has become increasingly clear that the solution to this disorder resides in combining various disciplines. The chaotic models of quantum physics may bring to light what governs genetic behavior. What I find most fascinating in this model is the concept of a network and the interaction of the components within the network. The more I evolve, the more I perceive reality as being a network of connections that may or may not interact in response to a specific frequency rate. Like quantum physics, which demonstrated through the superstring theory that matter can no longer be defined in terms of its chemical components, but rather in terms of the frequency and intensity of the vibrational rate of its elements, I perceive existence in terms of the frequency rate of a network of interacting and intersecting probabilities. The frequency rates of these various probabilities repel and attract one another, and we vibrate simultaneously in a network that continually changes according to the frequencies emitted.

Over the course of my thirty-some years of spiritual practice, I have felt a change of frequency on the planet. I am convinced that the time has finally come to transform the old paradigms upon which our genetic programs as humans are based. I am convinced that we are living in an era in which we need to reclaim the collective heritage that resides at the very core of each of the cells in our body. Everything that I know suggests that current planetary circumstances are favorable.

This current change of frequency involves a change in vibrational amplitude, which in turn affects the law of causality, whether on the side of darkness or of

 3 Ibid., p. 66.

light. On the one hand, we are now witnessing an increase in violence and world conflict. On the other hand, we observe that our intentions, when we voice them clearly, have almost instantaneous repercussions on an entire network.

We are akin to the pioneers who colonized North America. Weary of letting government authorities dictate their behavior, they had the courage to reinvent the world and themselves. We must not let go of our desire to modify, improve or redefine our programs. After 15 years of genetic research, science itself is losing its points of reference. Biologists are becoming acquainted with new dynamic models. "Biologists are beginning to believe that DNA will reveal its secrets through the dynamic description of its expression rather than through the static apprehension of its structure."[4]

How can we conceive of a dynamic model? The basis of dynamism is movement; it thus requires innovation. We need to set aside our cultural or scientific shackles and open ourselves to new discoveries. Who would have thought that genes could influence one another and that a gene could be "governed by several signals that can affect its expression in various ways, by increasing or diminishing its activity, or even by modifying it?"[5]

We already know that our attitudes influence our health and our evolution. We can also change the programs that govern us. We can become conscious participants, imaginative innovators.

4 Ibid., p. 65.
5 Ibid., p. 64.

CHAPTER 2
ZERO POINT

The time has come to reclaim our power and put an end to the polarization that comes with duality. This polarization implies that when I create solely from a perspective of light, I inevitably generate a negative force elsewhere. We are now at a stage of our evolution in which we must integrate these two polarities. To do so, we have to move beyond our biases for one polarity, be it the positive polarity (light) or the negative polarity (darkness). **The coexistence of opposing polarities is what I refer to as zero point,** a magnetic experience in which we choose to simultaneously integrate the negative and positive aspects of our experiences.

Tai chi is a good example of zero point. In tai chi, everything is fluidity and movement. Imagine for a moment the hands of a person doing tai chi and become immersed in their fluid and gracious dance. This is how we can imagine zero point: like two polarities in movement that communicate and are balanced without either one predominating. Imagine that one of the hands suddenly decides to take control and stops moving, or decides to move independently of its partner. Gone is the fluidity. Gone is the grace. The movements will automatically be unbalanced and, eventually, will stop. This is what happens when we are polarized: the hand that takes control, the predominant polarity, interrupts the choreography, and movement will not be restored until the two hands interact once again.

We all have moments in which we are paralyzed. This is caused by a default (unconscious) program and can happen in various situations. When we are polarized by shame or fear, we become paralyzed, and the only way to begin moving again is to unite shame or fear with its opposing polarity, for example, self-confidence. Even if we live in a polarized world, we can remain fluid through coexistence at zero point. We can combine our intentions and our commands and live in a state of love at zero point.

The state of love is beyond positive and negative polarities. In this space, referred as zero point, holy and evil, light and darkness, good and bad, as well as all other dualities, coexist in equilibrium. Zero point is neither neutral nor static. It is multidimensional and in constant movement. It is the center of a space that is in constant change because the positive force of light and the negative force of darkness coexist without canceling each other, despite their opposing polarities.

By balancing the two polarities within ourselves, we can choose to live in a state of compassion and love. If we manage to integrate the positive and negative charges of our experiences, we inevitably manifest a balanced and harmonious life. Because we have so often experienced the results of unbalanced polarities, I believe we are finally ready to integrate these polarities and create a new reality based on the state of love at zero point rather than on the energy of positive or negative polarities.

Outside of the state of love at zero point, we are polarized, whether in good or evil, in light or darkness, etc. But what happens when we carry out a task, a project or an activity in this state? We simultaneously manifest its opposite. Of course, our polarized actions can provide the expected results. But if we take a closer look, we will see that our project required more time, more energy and was more stressful than necessary. It is also possible that the end result, while acceptable, is not appropriate for our personal situation. But because this project is polarized, and therefore subject to the feedback law, it will automatically create its counterpart or opposing manifestation at the energetic level.

The only way to get out of this vicious circle is to accept the negative charges, use them and integrate them so that they become like the poles of a magnet that draw all that we need into our lives. It is difficult to maintain a very strong positive charge when the negative charge is low. It is even more difficult to manifest new realities if the magnetic force is too low and cannot *stick* to our lives like a magnet on a refrigerator.

The integration of negative charges is a powerful process. Imagine the intensity of an anger-based charge that has been accumulating for many years against an abusive parent. If we could measure the energy amplitude of this anger, we would be impressed by its magnetic charge. Once this charge has been integrated in our DNA, it becomes a strong negative pole to which we can match a corresponding positive pole without any need for anger. From now on, we will use these poles as our power base to manifest our desires.

The best way to envision zero point is to imagine a balloon full of water, floating in the middle of the ocean, spinning and rolling in all directions due to climatic changes and the motion of the waves. Zero point is located in the middle of this balloon. The core of this balloon is always balanced despite all outside turbulence.

Although I prefer the sphere-like image, it is also possible to imagine zero point as a fictitious point in the center of a straight line with a negative and positive polarity on either end. At zero point, the positive force is proven optimal because its potential is activated by the presence of the corresponding negative force, both charges being maintained in perfect balance. This being said, we can imagine that two opposing forces can coexist in the same space without ever having to join, just like the poles of a magnet.

Now that we will be using the magnetic charge of a once troubling and paralyzing emotion to manifest what we want at zero point, we will no longer see the negative charge as something to struggle against, but rather as a creative force. Here is an example of how to state an intention: "*I choose* to have fun, *even if* I am tired." To reinforce the idea that we incorporate all our emotions in our intention, we can sometimes replace the "*I choose… even if*" model by "*I choose… while*" (having fun, being afraid, doubting myself, hesitating, etc.). Regardless of how we state our intentions, what is important is that we create an intention while including every category of emotion. By doing so, we will feel new energy flowing and vibrating within. We will feel powerful. Conflicts will be settled very quickly because we will be taking advantage of them rather than trying to avoid them.

We often hear that we have to overcome our fear because just thinking of it can accentuate its power over us. It is also said that fear is so powerful that it can make the very thing that scares us come into being. It is precisely this power that we want to exploit at zero point. By allowing our fears to coexist with our hopes and our projects at zero point, we channel this power. When we include this strong emotion in our intentions, we provide a direction to the power generated by fear and use it to meet our needs.

Instead of fearing the disappointment that can follow a moment of happiness, we will now recognize and accept this feeling. We will then see to it that it serve our interests rather than harm us. Take, for example, a gathering such as a conference or a State summit. Have you noticed that even if they do everything to drive away alterglobalists, multinationals can never making them disappear completely? Driving away an energy that scares us or bothers us does not make it disappear.

One of my clients once said, "Every time I call light upon myself and I experience something extraordinary, it seems that everything that follows goes wrong!" I have had the same experience. Each time I had a truly extraordinary trip, I had a hard time coming home. Each time I had an extraordinary spiritual experience, I knew that I would have to integrate it afterwards through challenging "lessons" or by detoxifying myself or doing something of that nature. In other words, it seemed like "sunny days were always followed by rainy days." Now that I include my disappointment in these experiences, I no longer experience negative aftereffects. Another example of this phenomenon is the "yo-yo" effect that occurs when dieting. First there is the period of goodwill, when people deprive themselves to become thinner, feel good about themselves, or to clear their conscience. Then, several months later, in spite of all their efforts, they regain the lost weight. Diets are a perfect example of this polarization phenomenon that ends up causing stagnation and frustration rather than the expected dynamism.

Being at zero point does not mean letting things go. To the contrary, it involves maintaining two different polarities in the same space. We do not need to keep our weaknesses at arms length as though we were afraid of them. That's where we're wrong! They must be allowed to exist simultaneously with our good sides. Take the case of a client who was tormented by thoughts of meeting her ex-spouse in court for the divorce ruling. She had prepared herself well and had all the necessary documents in hand. She had developed solid arguments and was getting ready to present them with her head held high. But this perfect image of herself denied the fear, anger and hurt she had previously felt. Polarized in this way, her attitude presented an important weakness that her adversary would have the leisure of attacking. When she left the courthouse, my client was devastated and did not understand why. We can go back over this example and imagine that my client appears in court, well-prepared and proud of herself, but this time at zero point. No part of herself has been set aside, no aspect has been left out of her circle of strength or pulls her to one side or another. My client is focused, with healthy boundaries. Her positive and negative parts coexist in a circular whole. She is driven by a clear intention at zero point and appears before her adversary without any unacknowledged weaknesses. Her unshakeable certainty of winning coexists with her fear of failing. It is clear that the result of her encounter will be very different.

Another common misconception lies in believing that to be in the state of love at zero point means to have no limits and to be available to everyone's needs.

Someone once explained to me how important it was for her to devote herself to a cause and to be of service to humanity. However, she was frustrated by the frequent feeling of being taken advantage of by the very people she was trying to help. I answered that this was because her devotion was not at zero point. It is important to be at zero point in each and every decision that we make. If one day being at zero point means devoting oneself to others, so be it. The next day, zero point may be different because it has changed. Thus, our intention is not to choose to help or refrain from doing so. Our intention has a new focus, which is to live at zero point.

Now take the example of a woman whose mother, when pregnant with her, moved on several occasions and went through many upheavals. Now that she is an adult, this woman's life can be chaotic and disorderly in response to her early upbringing. She often feels confused and lost. Whenever she finds herself in this state, her old wound is reactivated. It is at a moment like this that she can choose to have chaos coexist with serenity. She can maintain these two opposites together through an intention at zero point, and experience this balanced state without having to change or erase her past. Being at zero point means knowing the limits we have as humans and welcoming them with love. Uniting light and darkness can produce results that are both extraordinary and powerful, but that are also comfortable and tolerated, because this unity is appropriate and generates love.

I have heard many astonishing stories about zero point. Here is one that was sent to me by a reader:

I was reading DNA Demystified *while waiting on the phone for the postal worker to give me information regarding a parcel for which I was going to have to pay $45, even if I decided to return it to the sender. The postal worker stated that she could do nothing, that there were new rules, etc. With unshakeable calm, I told her that I did not accept that answer. She then transferred my call to her superior. I chose to put the situation at zero point while accepting my doubts. At first, the superior accepted to reimburse the $45. As we spoke, we realized that I had to pay $45 to obtain the parcel and $45 more to send it back. The woman therefore decided to mail me a check for $90! This zero point business really works!*

Morele D.

Those of my clients who have worked with zero point often speak of a state of ease and well-being that requires very little effort. The more we choose to be at zero point, the more we will vary the scope of our experiences and experience new emotional states. This way of thinking includes more freedom and a sense of love. The rainbow of experiences that colors our world at zero point is much more diversified than the black, white and gray to which we have had access up to now! Suddenly, we are accessing new data, having new experiences, and achieving greater things.

You too will experience a feeling of fullness when you allow yourself to feel both irritation and pleasure, with these two experiences remaining whole and not merging into something neutral. You may experience some difficulties at first, because it is like holding a positive charge in one hand and a negative charge in the other, with these two charges repelling each other. The key lies in viewing this experience as a laboratory experiment and observing what happens **when you intentionally choose to be at zero point.** Suddenly, the tension within dissolves and an unusual but harmonious energy springs up within you. By simply allowing your two emotions to coexist in the same space like the two poles of a magnet, you will create a magnetic field. You will be able to feel this field. Just command zero point and wait a little. The field you will feel as the result of the coexistence of two opposing forces is the vibration of zero point.

We will try this out by selecting any given fear, such as the fear of never feeling secure because we might think we do not deserve safety. We will allow ourselves to feel this fear while resisting the temptation to repress it. However restricted we may feel, we will be able to simultaneously experience the vibration of expansion and safety. Darkness can vibrate in the presence of light, and light can vibrate in the presence of darkness. We maintain the tension of these opposing forces until we feel that our limits have dissolved and we feel a greater well-being than if we were polarized.

How can we be at zero point, or how can we approach it steadily and constantly? We do so by using the power of intention. For example, when you wake up each morning, you could state the following intention: "Today, *I choose* to be at zero point *even if* I don't know how." You can also practice putting all the events of your daily life at zero point. The car mechanic is getting on your nerves? Change your frequency by saying internally, "*I choose* to vibrate at another frequency *while* feeling frustrated by the mechanic's attitude." You are waiting in line in a store and start to flip your lid? Say, "*I choose* to put this experience at zero point *while* feeling irritated by the cashier's slowness."

By beginning with these small daily events, you will accumulate experiences at zero point. Once you have accumulated enough of them, you will discover a new way of life, and, rather than being governed by old programs, you will start to have experiences that are more original.

CHAPTER 3
THE CHOICE OF A POSSIBLE REALITY

We have a lot more power than we think. You have surely noticed that it seems easier for certain people to manifest their desires. When they have their sights on an objective, they do not find themselves up against endless obstacles and do not meet with rejection upon rejection. Their actions produce quick results with a minimum of effort. We do not have to be special to do the same. In fact, these people use the power of intention very effectively, but they mostly do it without being aware of it.

Intentions shape reality by giving it a particular vibration. The problem with a stagnant situation, or with a project that seems to take forever before showing any result, is that it is not taking place on the right quantum path. The power of intention allows us to redirect the situation along a path that is in harmony with the frequency rate of our essence. It is then given the proper momentum.

According to the principles of quantum physics, each observable event was initially one of a series of possible choices. The observer's choice made it possible for a particular possibility to materialize. All the other possibilities, those that were not chosen, continue to exist in parallel universes. Once the decision has been made, we embark upon an infinite sequence of causes and effects. Each of the latent possibilities possesses its own frequency rate.

In the world of the infinitesimal, a physicist assesses the probabilities of a particle's actions using highly complex calculations. He determines its spectrum of activity by comparing its previous behavior and then decides what the most probable behavior will be. At the exact moment the decision is made, the nature of the particle is changed. Those people whose intentions have great impact have a very clear understanding of what they want and who they are. **Their understanding is in tune with the frequency rate of their essence.** These people are able to attract what they desire because they are **magnetic**. We will therefore explore certain techniques that will allow us to clarify what we really want.

Take Lucy, for example, a flight attendant. Many crew members never get the schedules and destinations they want, unlike some of their colleagues who seem to be able to do so. Lucy decided that she wanted to work on transatlantic flights. She therefore went to see her superior to request to be assigned these flights. She prepared for the interview, mentally coming up with several scenarios. In the end, her superior informed her that her name had been placed on a waiting list because the company had made a lot of cutbacks. Lucy will just have to put up with the delay.

One month later, however, one of her colleagues made the same request and was transferred the following week. What happened? Her colleague possessed, no doubt unconsciously, a secret that we will decode here and that has to do with the magnetic power of conscious intention.

The Magnetic Intention

To understand how to use intentions, we must first leave behind wishful thinking as well as our old programs and, instead, become our own programmer. We will then use our doubts and weaknesses as a negative pole, and our new choices as a positive pole. This will allow us to become magnetic and command the actualization of new possibilities.

In this context, an intention is a command voiced out loud concerning the state we wish to attain. Voicing an intention is an act through which we **take responsibility** and regain control over our lives. An intention is therefore a sentence (a command) that describes both the new program to be installed as well as its unconscious negative counterpart that could activate itself **by default.**

To create an effective magnetic intention, we first have to identify our suffering and our weakness or, in other words, the state in which we are most inclined to activate our old programs. Next, we have to include this state in the wording of our intention.

Just as the action of a magnet stems from the presence of two poles, positive and negative, our intentions or commands must also have two poles corresponding to the two aspects of the equation. For example, saying, "*I choose* to obtain an appropriate schedule and itinerary" only becomes a powerful intention when we add "*while* doubting that this is possible because of cutbacks in positions." When we integrate these two poles, our conscious intention is magnetized and becomes effective.

Formulating an intention based on the binary model is a good way to ground it in daily life. If I manage to stay in touch with my discomfort–without wanting to get rid of it–and simply seek to acknowledge its presence and understand it, I will then be able to use it as the negative pole of my intention. So, instead of feeling negative or powerless, I will be able to use my vulnerability as the negative pole of my intention. I will integrate it into my command without judging it. Consequently, I free myself from a polarized way of thinking and I begin a process that is not linear, but rather **circular** or global, and which belongs to the quantum dimension.

What precedes illustrates well the fact that our **salvation**, our real power, comes from accepting our humanity rather than rejecting it. We tend to magnify our divine side and tell ourselves, "I am beautiful only when I show my bright side." True beauty, true power, comes from a balance between our light and our darkness, the latter being our more fragile and vulnerable side.

The Insidious Effect of Wishful Thinking and Expectations

Wishful thinking is one of the reasons we allow ourselves to be led by outmoded, by-default intentions. This expression refers to the habit we have of not paying attention to things, of not being bothered, of thinking that problems will sort themselves out in time. Wishful thinking is believing that one day everything will turn out okay without doing any inner work. That you will miraculously win a million dollars, or that all of a sudden the world will discover that you are a talented artist. In other words, it is believing that all of your wishes will come true as if by magic.

The first time I stopped to consider the meaning of the expression "wishful thinking" I really had to take an objective look at myself to fully comprehend how this concept applied to my own life. I realized that I had been hoping that, on some hypothetical day in the future, by sheer chemistry, I would have the life that would make me happy. I had not yet understood that I would have to stick to the task and come to terms with my vulnerabilities, rather than pushing them away, hoping that everything would resolve itself. I had many ups and downs before I was finally responsible enough to face up to the reality of my life.

From 1980 to 1990, wishful thinking was frequently manifested in the form of affirmations. We thought that all we needed to do to improve our lives was to make positive affirmations, such as the classic, "Each day, in every way, I feel better and better." I used those affirmations often. However, I have to admit that they did not help me overcome the challenges that I had to deal with on a daily basis. Faced with the "catch-22" situations of everyday life and the constraints of the third dimension, affirmations seemed to create more frustration than contentment in my life. And yet, these affirmations were intentions or commands in their own right, were they not?

Wishful thinking convinces us that everything should be simple, and consequently, we often get discouraged when it is not. It's at the root of the complaint we often hear in therapy, "Oh no, not that story again! I thought I'd already dealt with that, that I was done with it."

When we hope to see our discomfort "magically" disappear, we abdicate our role as programmers, relinquishing our power over our reality. Nonetheless, life requires that we get involved in our evolution. It wants us to be imaginative co-creators who work with **conscious intentions** *while* granting a proper place to our sufferings. And therein lies the key, in the words "*while* granting a proper place to our sufferings." It took several years for me to understand that default programs were canceling the effect of my affirmations because I was in the habit of denying my pain and vulnerability, in other words, because I was denying my humanity and duality.

There is another form of wishful thinking that hampers our progression at zero point. It has to do with our unrealistic expectations and our desires, or with assumptions and possible realities that get our imagination moving. This type of wishful thinking takes the form of scenarios or desires made up of publicity images in which we are stars with well-filled bank accounts and romantic relationships that require no effort. These desires are not needs. These are the aspirations of teenagers who refuse to grow up. Take, for example, one of my clients who found it unfair that his spouse did not have to work because she had received an inheritance following the death of her parents whom she had loved deeply, while he had to work to earn his daily bread. He slowly fell into depression, convinced that his life was a failure because his own path was different from that of his wife. It became absolutely essential that this man emerge from his trance and understand that each person has a unique life path.

By choosing to free ourselves from unrealistic imaginings that are based upon assumptions, we will finally know a peace even greater than that promised us by wishful thinking. By becoming enthusiasts of reality at zero point, and with a bit of patience, we will find ourselves with roles in new scenarios and we will experience new emotions. We can now use the hidden power of the disappointments that these unrealistic assumptions have caused us to command situations aligned on our reality at zero point.

These assumptions leave us in limbo, waiting for a hypothetical future, rather than rooting us in the present. They are reflections of a reality that does not necessarily vibrate at our own frequency and that keeps us in inertia.

Recognizing the Duality of an Intention

I CHOOSE... WHILE...

With the formula "I choose... while..." the intention **sticks** like a magnet on a refrigerator door. The first part (the one that follows *I choose*) expresses the new program that we wish to install, whereas the second part (the one that follows *while*) expresses the weakness or unconscious (default) program that we wish to replace. When you word your intentions in this way, you will make them effective and magnetic, and you will undoubtedly notice their effects in your daily life.

Unconsciously, we all more or less feel the phenomenal power of our wounds. Our fears, if they are nourished, actually have the power to manifest what we fear the most. By becoming conscious of the creative potential of our fears, we can make them the driving force of our commands and choose new avenues that correspond to who we are. If we do not direct them consciously, they will quite naturally follow the course of the default programs that emanate from our deepest wounds. From now on, we command that our fear coexist with a feeling of safety. By having these two emotions coexist like the poles of a magnet, they can reach an equivalent magnitude. Our feeling of safety then has a force equivalent to that of our fear.

In my opinion, the "I choose... while..." formula is also the only way we will succeed in loving all aspects of ourselves. Of course, it is not always easy to accept our vulnerability, our sufferings and our weaknesses. A magnetic intention, like the following, enables us to do so: "*I choose* to love myself *while* accepting the unpleasant situation I am currently experiencing." This is why it is so important to remain in touch with our weaknesses and use them to manifest a new reality in our lives. If we take the time to properly define the new program that we wish to install, we will succeed in integrating our dark sides into a new paradigm free of binary restrictions.

If you do not know how to state your intentions, or doubt your power to do so, here are a few that have been very useful to me and that you can adapt to your situation. They will thus become your new default intentions:

"*I choose* to love myself *while* criticizing myself and/or lacking self-love."

"*I choose* that this work be easy and effortless *while* doubting my competence in this field."

"*I choose* to be happy *even if* I don't know how."

"*I choose* to trust my body *even though* I am suffering and feeling sick."

"*I choose* to be competent *even though* I am lacking self-confidence."

"*I choose* for everyone to win in this conflict *even though* I am feeling stubborn."

Even if we include our duality and focus on the state we aspire to, it is important to understand that we can only use intentions that respect others. Therefore, it is not recommended to state our intentions in terms of one person in particular. For example, rather than saying, "*I choose* that these individuals not be able to get in touch with me *even though* I am afraid of them," it is better to focus on our feeling of safety by stating: "*I choose* to feel safe, loved and respected *while* being afraid." Instead of voicing the intention to find a rich and powerful spouse, we could *choose* to feel protected in an intimate relationship and to have abundance *while* feeling distrust, for example. If I was ignored and isolated when I was young, instead of saying, "*I choose* to have a very attentive husband," I can *choose* to feel important in an intimate relationship *while* feeling resistance.

The State/Cause Before the Form/Effect

By using the power of intention on a daily basis, I have come to understand a concept that I consistently apply and which strengthens my intentions even more. This concept has to do with the fact that once an intention has been stated, we have to release all expectations about the form in which it will manifest itself. If not, we will lose much energy trying to control the result rather than welcoming its effect. Working with the state/cause rather than the form/effect leads us to a new paradigm beyond the limits created by our past experiences. Of course, we have to include our personal needs while we are describing the state we wish to experience. It is only the eventual expression of its form that we have to release. For example, I could decide that when I experience intimacy (not naming any one in particular), *I choose* that this person vibrate at the same frequency rate as mine *even if* I do not know how to make this happen. I must then trust the power of my intention. This magnetic power will draw someone to me, anyone, who will be aligned perfectly with my frequency rate, but who will not necessarily have the appearance or the style I would have imagined.

There is quite a difference between "*I choose* to be in love with someone rich, famous, beautiful, etc." (emphasis on the **form**) and "*I choose* to live in the state of love *even though* I am afraid of the challenge of intimacy" (emphasis on the **state**). In the first case, failure is more than probable and will trigger a feeling of depression, and the impression that we are a victim or unlucky. The second formulation emphasizes the state that we are looking for (the state of love). Everyone knows that when we do not love ourselves, no one is interested in us and that, conversely, many **good matches** present themselves simultaneously when we fall in love. It is simply because we are already in the **state** of love and have stopped looking for its **form**.

We all feel powerless at some point or another. Imagine that everyone around you has the flu. Since you have not contracted it, you congratulate yourself on the effectiveness of your intentions. But then, you fall sick! Right away you search for the cause of this state and you criticize yourself for having formulated intentions that were too weak, for having made an error, etc. THIS IS THE PERFECT MOMENT to put these sad thoughts to good use and to affirm that you choose to be healthy *even though* you feel unworthy or not good enough. Try it! You will see that your period of convalescence will be much shortened. It is in these moments that we need to trust our intentions rather than losing ourselves in a labyrinth of questions and strategies. When in doubt, we have to remind ourselves that we have *stated an intention* and continue to trust it *while having doubts.*

The intimacy of emotional relationships is a very favorable environment for practicing conscious intentions. Instead of reacting by criticizing our spouse, which solves nothing at all, we can work with magnetic intentions. Thus, as soon as a conflict arises, we can choose that our couple be protected while accepting the strength of our anger. Similarly, here is a useful intention when communication with our partner is blocked or difficult: "*I choose* to communicate easily in my intimate relationships *even if* I do not know how" (or *while* having difficulty expressing myself because my father was very domineering, etc.).

Stating a conscious intention each time a conflict arises with my spouse has resulted in new situations in our relationship. WE MUST NOT BECOME DISCOURAGED! Even if we still experience conflicts, I have noticed that they have less of an impact and do not weaken our relationship. I feel that our relationship is increasingly genuine and that we are finally beginning to emerge from the infernal cycle of projection and blame.

One of my clients–we will call her Jennifer–wanted to have a house in the country. She first had to identify the type of house she wanted. A good way to do this is to create a collage. She therefore collected a large number of magazines. Next, she and her husband chose a quiet moment to flip through the magazines to the sound of soft music. This step calls on the right brain–our intuitive brain. Without analyzing or using scissors–a left brain activity–they retained all the images or words that pleased them, without questioning their motives. The next phase consisted of cutting out these images and gluing them on a large piece of cardboard. This step calls on the left hemisphere of the brain, which is more rational. The first step is rapid and intuitive whereas the second can last much longer. To understand our priorities, it is relevant to observe which images we glue in the center and which ones we glue further from the center. This little trick enables us to clearly see our needs. Jennifer could just as well have drawn up a list, read tarot cards, or referred to a list of emotions. In fact, the important thing is to let inspiration and desire rise within us without judging.

Now that Jennifer carries the image of her future house inside herself, she can look at the *yes, buts*. These objections will serve as the negative pole. She has to take the time to feel them properly if she wants to obtain a productive pole. Now that she and her husband carry the image of their house within themselves, now that they have clarified what they want and have allowed their doubts to emerge, they can *state together the intention* to easily find the perfect house for themselves *while* feeling hesitant and having doubts. The house thus took shape within before manifesting itself externally. In the end, the couple found the perfect house; it required no renovations and it happened effortlessly.

My work as a coach has allowed me to collect dozens of similar stories. Thanks to this method, we are no longer manipulated by events; rather, the latter bend to our intentions.

To progress, we absolutely have to choose an intention that is of utmost importance to us. I often think that the problem is not in formulating the intention but in working out its content. It is essential not only to understand our personal needs, like Jennifer and her husband did, but also to strongly desire that our intention become reality. If I choose to ask for a house that is ready for immediate possession, but deep within I am willing to compromise, there is a good chance the compromises will prevail. Rather than having to make compromises, LET US DARE to choose and use our compromising doubts as a negative pole so that our intentions are magnetic. We have to understand how to command what we want rather than what we are afraid of.

When magnetic individuals enter a room, they seem to attract people. People gather around them like moths around a flame. These individuals receive tribute and proposals without lifting a finger. We too can become magnetic by using the magic wand of magnetic intentions at zero point.

Effective Commands

Quantum intelligence invites us to adopt a circular way of thinking and to manifest abundance by accepting our negativity rather than being defensive and seeking to escape the intensity of modern life. We have to continue to create and to use our free will by commanding the creativity **associated** with a quality of life that reflects who we are.

This new paradigm is our only solution in the face of the despair affecting humanity at this time. We can use the magnetic force of our despair to manifest a future that is blessed rather than give up and renounce our role of creator.

To do so, we need to think in terms of quantum physics. Imagine a blackboard marked with parallel lines. These lines represent realities that vibrate simultaneously at various frequencies in parallel worlds. We can choose from these realities the one in which we wish to live. Imagine that the despair related to the environmental degradation of our planet is represented by a vivid red line on this blackboard. Because this vivid red line is most prominent, we feel obliged to choose it. If this line is predominant, it is only because a large number of people continually go over it again and again with red chalk by choosing the reality that the media, among others, conveys daily.

And yet this blackboard contains other lines, lemon yellow or pistachio green. Every time someone chooses a green chalk, the pistachio green line becomes darker. The choices of the masses influence our reality. The more people there are who choose to believe in something or who adopt a particular habit, the more this belief or habit becomes grounded in daily life. These are the resonance principles of the 3-D universe in which we live. In the face of global despair, the same laws apply. We are convinced that we need to be hyperachievers and make progress, otherwise failure lies in wait. We are involved in a wild race to save ourselves from the despair represented by the red line on the blackboard.

Yet, we can change quantum paths by establishing **healthy boundaries**, by becoming aware of our **personal and authentic needs**, by using the power of **intention at zero point** and, finally, by making the most of the extraordinary power of the **frequency rate of our individual and unique essence.**

Take the case of the automobile. Our car is a source of considerable pollution at all stages of its useful life: when it is built, while it is being used and when it is sent to the dump where it contaminates the earth. On the one hand, when we use it we may feel guilty and powerless. On the other hand, if we cannot do otherwise, we can decide to reinforce another line on the blackboard by using the strength of our despair and our feeling of powerlessness to command that the planet and the environment be protected. The pollution problem is directly related to the weakness of our intentions regarding the protection of our planet. We have no connection with our planet. We constantly go back over the vivid red line instead of supporting another reality that would automatically and effortlessly lead us to adopt living habits that correspond to our vision. We should therefore continue to manifest a reality of safety and love for our mother the Earth, rather than be victims of the consequences of the reality corresponding to the red line. We will then become creators. The more we make conscious choices, the more we will identify with our role as creators.

We have good reasons for believing that we cannot be creators. These reasons have enormous power and we can use this power for our commands. And yet, instead of using this force of resistance to our advantage, we polarize and criticize ourselves. We believe that the more we change our consumption patterns, the healthier the Earth will become. When all is said and done, the situation is only worsening. To choose a code of conduct, we need to become aware of the fact that a frequency cannot be supported by the form/effect. It is the state to which we aspire that has to be in control. By maintaining the frequency of a safe Earth, we add a stroke to our 'green' choice on our blackboard instead of reinforcing the old paradigm. When we fight the reality of the red line instead of changing paths, we only reinforce this red reality. To thwart its effects, we need a chalk of another color.

We tend to justify ourselves by saying that we are powerless in the face of a global context that is more powerful than we are. Each time we choose to think in this way, we draw the chalk along the red line on the blackboard.

The solution is not external–the attitude we need to change is our feeling of powerlessness. The more there are people who direct their lives by using the power of intention, the more rays of different colors there will be on our blackboard and the vivid red line of global despair will become less predominant. Through our individual and respective decisions, we will amplify the frequency rate of free will and our blackboard will contain all sorts of nuances that will coexist instead of being dominated by a single frequency rate of despair.

For example, instead of thinking that we need a large community to feel that we belong, we can choose to feel that we belong even while isolated in a rather closed environment. We can choose to command what seems appropriate for us instead of choosing the reality of the society around us. After a few months, we may come to realize that even though we only spend time with a few friends or another family, they are our community, this community is solid and we truly feel that we belong. By commanding states of being, such as a community feeling, rather than waiting for things to happen, we stop feeling powerless. Because if I wait for one of my friends to be less busy to feel that I belong, I risk feeling frustrated. On the other hand, *if I voice the intention* of feeling that I belong, *while* embracing my uncertainty, my fear, my disappointment and my inertia, my desire will manifest itself because I have maintained a clear intention. Instead of imagining the utopia I will one day experience within a large spiritual and ecological community, I can start feeling that I belong to a community of three or four people whose frequency is in resonance with mine.

People who aspire to finding a better job have to choose to be happy and satisfied at work, and to get along well with their colleagues. From the moment that they vibrate at the frequency rate of this intention, they will be able to manifest a job that corresponds to this frequency. We seem to be waiting for the government, God or someone more important to fill our needs instead of manifesting the state that we want and commanding the feeling we will experience in this state. If we manifest the state, the form will inevitably follow.

I know people who pray and who achieve good results. In my opinion, an intention and a prayer are pretty much the same thing, since both are a type of command. Whether we formulate an intention or a prayer, we need to do so at zero point and include our doubts. What prevents people from praying or commanding what they want is that they doubt themselves or the result of their prayers or commands. It is important that people include their doubts in their commands or their prayers because these doubts constitute one of their magnetic poles.

There are individuals who believe in prayer without being religious. They recite a new type of prayer addressed to the Source of the universe at zero point. These people are convinced that their prayers will be answered because they include their doubts and the negative magnetic charge of a lack of personal worth in their prayers. In my opinion, we have to simply include these feelings, not do away with them. If I say every morning, "*I choose* to move forward in my life *while* accepting my resistance," the form/effect of this command will manifest itself. We are convinced that we have to become less rigid, get rid of our transgenerational memories, or stop feeling physical discomfort to be able to command. On the contrary, we have to include these elements in our commands. This requires us to clarify what we really want and this is often the hard part.

This leads me to the question of conscious simplicity. People may use this expression to avoid having to ask themselves what they want to manifest. In fact, they do it to refrain from becoming creators because, when we do not command, we do not create and we do not participate in co-creation. We are humans living the human experience on the planet Earth. We are supposed to co-create; we are gods incarnated here to know what it is to be human. We are not supposed to feel powerless in this experience. Interestingly, when we finally realize this, we feel a pleasure that is directly proportional with the height of the barrier that previously prevented us from moving forward. We are then filled with an exponential power that encompasses everything.

And to think that we have erected our barriers precisely to achieve this state of power! Yet, we are convinced that we will never manage to tear down these barriers. Rather, we can affirm that "We do not want to get rid of our barriers. It is with the power of these barriers that we wish to command, because zero point is the new paradigm." Zero point lies outside of duality and this is the reason why we have been incarnated on Earth. To become experts in duality, we need to let go of the emotional memory of a lost paradise where we did not have to command specific things. I would say that people who adopt the polarized model of conscious simplicity do not assume responsibility for manifesting. People believe that one day they will find a paradise where everything will be perfect and where other people will make decisions for them... In the Garden of Eden, there were no decisions to be made, no conflicts to resolve, no desires to satisfy, nothing. We have come here to have desires, to overcome obstacles–that is what the human experience is about. Basically, we should say, "Today *I choose* to fully live the human experience, *even if* I don't know how."

To feel that I have enough worth to command, I have to appreciate certain aspects of myself. I have to decide that I have worth, that my essence is unique and have faith in my uniqueness. Even if we feel that we are being increasingly controlled and are losing certain liberties, our vital energy in itself is not affected. The energy that causes us to breathe–breathing itself–does not depend on anyone. Within us lies a power that is greater than certainty and that power is confidence. We need to have confidence in the energy of who we are... I may want my team to be competent; I may want to plan beautiful projects; I may want to feel that I belong; I may want to feel creative. Our power of manifestation, however, can only be exercised in respect to one thing alone: ourselves. The greatest act of shamanism is not to connect ourselves to energy, it is to connect ourselves to ourselves and to command that our daily lives be joyful and full of life. In my eyes, shamanism only has worth if it has repercussions in our daily lives.

Rather than saying *I command a specific job*, it is better to say *I command a suitable job for myself.* If it is the one I have my sights on, great! If not, I want the equivalent or better." When, or if, you worry about whether your command is effective, just repeat to yourself: "My command has been given." Globally, we are now moving to a stage where all of our commands will be in tune with our specific frequencies. *I command* the state of love in my job *while* being dissatisfied with the lack of integrity of my boss or my colleagues. I am human. I have been incarnated here on Earth. I am a pillar of frequency. Wherever I am, I want to live in the frequency rate of love *even if* I don't know how.

If we wait for the instruction book to show us how to make commands, we will wait too long. When we do not command, the commands from the community and from the media are transferred and they become predominant. It is very unlikely that the media command love. Our worth, the phase in which we find ourselves, makes the Earth a better place. There are currently many **hawks** who hold positions of power throughout the world. This does not mean that the Earth is only populated by warriors. We may have the impression that we have no power because our actions have a very limited scope compared to those of multinationals, but when a large number of people become aligned on the same frequency rate, their power becomes exponential.

Hope is one of our strongest drives, and the stimulation that comes from hope can become a driving force for our commands. When you feel destabilized, command clarity. In our commands, we need to make the most of an emotional trigger. We must not let go of our dreams because we are angry or incapable of forgiving the wrongs done to us. Rather, we should use our grudges as the driving force of our commands. Dreams and wishes are meant to be reframed based on the reality of the moment. Remember that the energy generated by our emotions does not change and that we have the right to manifest our dreams and our worth in our lives, *even if* we are not perfect. Our commands have to be in keeping with our experiences and our desires. The person who has had a very chaotic childhood has the right to command calm and clarity. Someone who has been exiled as a child has the right to command roots. Without denying our childhood, we need to determine our future choices in keeping with our degree of maturity. The power of our choices will be increased tenfold by including what happens to us, stimulates us or destabilizes us.

Our commands must not address anyone in particular. A father cannot support his child emotionally if he never received such support himself. However, children-become-adults can understand that their fathers were unable to give them what they did not have. These individuals can then command support at large from the universe. They have the right to command what they need, despite the shortcomings of their childhood. The universe is plentiful and the support that these people need can come from elsewhere.

I understand that certain people can feel deep resentment towards a parent who did not meet their needs and that this leads them to deprive themselves of support or help. However, there are cases where we can let go and forgive our parents, knowing that they are people who were lacking themselves, and yet we do not have to accept what we went through. We can criticize the act while understanding the weakness of its author.

Imagine that a person married a distracted man. She could hold his many oversights against him daily. But she could also use the energetic intensity of her grudge to command that she be supported while feeling frustrated. She could criticize her husband's distracted attitude while forgiving him. It is possible to forgive without accepting the situation for all that. Perhaps her husband had a very demanding mother and the only way for him to survive may have been to develop a distracted attitude. By using her frustration to command support and to ease her life, his spouse could take advantage of the situation to free herself rather than becoming stuck in a polarized position and embodying in her turn the demanding *Mom*. LET US USE THE WEIGHT OF OUR FRUSTRATIONS IN OUR COMMANDS instead of being ashamed of them or being cornered in a dead end that is transmitted from generation to generation, because that will get us nowhere… We know that well enough.

Take the example of Judy, who outright refused to work overtime hours at the hospital, but who was continually assigned exacting schedules. Judy had to go back and feel the frustration she had felt as a baby when her mother did not have time to take care of her. After providing herself with comfort and support, she was able to speak firmly to her superior and tell him that she would no longer work overtime hours.

If she appears before her superior without being convinced of her personal worth, it is almost certain that her request will not be heard. But if she appears before him with the conviction that even if she was not held in esteem when she was small, she can make a different choice in her adult life, she can use this feeling of powerlessness as the negative pole of her magnet and command a schedule that suits her. Next, she will have to let go and wait to see how her wish manifests itself.

In a boundless universe that holds billions of possibilities, Judy's wish will manifest itself in accordance with what she commanded. We never know in advance what form the fulfillment of our wishes will take, but we know that this form will be in keeping with our intention–the universe is our accomplice and it is as though it were winking at us!

Imagine that at the end of the day the hospital places an assistant at Judy's disposal. She will go to sleep feeling conscious of her worth and loving her neighbor and life instead of feeling caught. Our feeling of being trapped stems from the fact that we do not command what we want. Judy had *voiced the intention* of taking time for herself this year and of being conscious of her worth *even if* she didn't know how. We should not fall victim to our desire to heal, but position our healing in the direction of our desires. I have come to realize that most people are so little programmed for success that they have great difficulty determining what they want. They have difficulty affirming which positive things they would like to see in their lives.

Here is another example of someone who was beaten when she was small and who constantly reacts as though she were being beaten again. Every strong emotion feels like a slap. The least bit of stress resembles a clout, even if this person is vigilant and watchful of her behavior. She has to come to grips with the stress she underwent as a child. If she attracts stressful situations, it is because the little girl inside her has been left alone with her memories. Instead of creating stress in our lives, we need to seek out the inner child who went through these stressful states. This child also felt powerless and her only solution was dissociation. Instead of repeating to her that it is over and she should be over it, you might tell your inner child, "I am there for you. You are no longer alone. There is an adult here with you and she does not approve of you being hit." You are then working with presence and intention. "*I choose* to live in safety *while* integrating the insecurity I went through in my childhood when I was not protected." In fact, what you want is to be protected. You can torture your mind to discover why you were not protected when you were young. Or you could tell yourself, for example: "this position of vulnerability has set off a powerful emotion in me; I will now use the energy from this emotion to fight my cancer and restore my health."

People who are looking for work have to cultivate a broad understanding. They must draw up a list of the states that they want to feel rather than the characteristics of their job. For example, one person might want a job where there are few emergencies, where the atmosphere is cheerful and relaxed, and where she feels that she wields a certain power and is in the right place. If you have trouble coming up with ideas, you can make a collage, use tarot cards or run through a list of emotions. Ask yourself what you want to eliminate, correct and possess. Then, have faith. Instead of focusing your attention outside, concentrate on the beauty of your essence, on what is beautiful inside yourself. When we are magnetic, we not only attract negative charges but positive ones as well!

CHAPTRE 4
DEFAULT PROGRAMS

We may be born with innate programs but the experiences we acquire influence our reactions. An article published in *Science et Vie* in April 2005 reported scientific proof that our emotional relationships can change our genetic code: "...geneticists and development specialists have discovered that significant chemical changes appear in the DNA of newborns... as soon as they come into contact with their mothers. In other words, cuddling has an impact that is not so much psychological as... genetic! (...) Thus, a mother's caresses can 'activate' genes located inside the neurons of the hippocampus. (...) It has been proven that the environment has as much influence on us as our heredity." [6]

The frequency rates of our default programs–or unconscious programs–are not necessarily in harmony with our own frequency rate. And yet, we generally base our decisions on these programs. When we were children, we integrated family programs that seemed perfect because they came from our parents. In fact, most of our default programs come from our childhood. They consist of nonverbal orders that the adults in our family circle gave us and that we integrated. Thus, someone who learned as a child that to be loved she had to be quiet and not make a sound, will tend to automatically chose the **silent method** whenever she feels the need to be loved, because such is her default program.

A child's brain and programs can be easily influenced (or programmed) because children tend to imitate their parents and those around them to feel that they belong. Moreover, this capacity for imitation is what allows us to learn our mother tongue. To be loved, we have all chosen programs, which, from a child's point of view, once seemed appropriate. The problem arises when we become adults, since these deep-rooted and genetically-encoded programs are carried on by default.

A childhood wound can have resulted in such deep trauma that it has created a sub-personality that is far removed from our essence. Take Laura, for example, who was brought up by a mother who suffered great resentment and envied her daughter. Her mother belittled each of her successes and cut off her drive. She now experiences the same scenario with her spouse: each time she succeeds at something, her spouse reacts like her mother would have in the past. This is hardly surprising because Laura learned at a very young age that pleasure is not healthy.

 6 Ibid., p. 90-92 .

The imprint of this wound can be transmitted from generation to generation. Laura's mother had lived in an orphanage and had learned to be constantly on the alert for unkindness from certain sisters. She therefore transmitted this behavior to her daughter. But since her daughter sees things from another angle, she can break away from this psychological legacy.

Some incidents in our adult lives are so like those of our childhood that they act like hypnotic inductors and bring us back into the past. It may be a tone of voice, an attitude or even the wallpaper that reminds us of our childhood bedroom. Something is triggered. When Laura tries to share the joy of her success with her husband, she falls back into the past. Her hypothalamus begins to secrete peptides and she once again becomes a little victim talking to her Mommy. As a result, she expects her pleasure to be spoiled. This is the time for an intention at zero point. It will allow Laura to change this pattern. She can make a new choice: "*I choose* to trust *while* feeling anxiety." She can change her perceptions and her possibilities even if she is convinced that she will always obtain the same negative reaction when she tries to share her successes with someone. Happiness for Laura is akin to disobeying her mother. Eventually, by becoming aware of this acquired model and by using an intention at zero point, Laura could free herself from it.

We Do Not Need To Change

We carry our entire past within. We cannot rid ourselves of our childhood and our family is ours forever. Our childhood wounds are there to stay. When we accept our life without judging it, we see that it is useless to want to change ourselves; it is our intention that has to change. True compassion consists of accepting the imperfection of our lives and then moving on to the phase where we can state our intentions at zero point.

To create powerful intentions, Laura must not try to change, she simply needs to choose another possibility that differs from one that her mother taught her. She can select a new state. She can *choose* to feel pleasure *while* accepting her hesitations. Instead of looking for a partner who will be able to make her happy, she can cultivate, on a daily basis, a state in which she is happy *even though* she is expecting punishment, etc. By commanding this state day after day, she will come to develop a feeling of confidence and will automatically find a spouse who will support her enthusiasm. If she is already in a relationship, she will observe that her clear intention will transform it.

Learning to observe our life, without judging it, will allow us to get a better understanding of the transgenerational patterns we have inherited. We will also understand that, owing to what our parents have given and taught us, and what we have learned in our environment, we may have failed some of our experiences. This does not mean that the information from these experiences is not currently available in the universe! Because we did not receive encouragement from a parent does not mean that support does not exist in the universe.

We can be daring and imagine different scxenarios that will allow us to make new and more powerful choices than those that we learned in childhood. We do not need to be heroic to become available to ourselves. We simply have to use our imagination and not expect perfection! It is OK to do this one step at a time. We can start by remaining present to our small aches and pains. For example, try waiting a bit before taking a pill when you have a headache. Breathe while in pain, without judging yourself. Take the time to feel the pain. The key to compassion lies in the capacity to remain present to ourselves without judging.

The Magnetic Frequency Rate of a Wound

Any danger or threat that we experienced as a child holds us in its grip because the wound is electromagnetic and continues to respond to our experiences. It continuously seeks out other frequencies in resonance with its own. It leads us toward people who resonate at the same frequency rate. In reality, everything is vibration and wounds also have very distinctive frequency waves that have the power to set off biochemical responses in the brain.

This vibrational attraction influences our intimate relationships. I have noticed that partners attract each other **by** the best and **by** the worst. When a conflict sets two people against each other, they communicate through their wounds, because they are similar. If we understand the nature of our common wound, we will be able to establish a new problem-solving protocol. We are attracted by people who carry the same types of wounds we do. Being electromagnetic, the wounds set each other off electrically. When one of the partners falls back into childhood, this trance-like state emits a vibrational pulse that sets off a ripple reaction in the brain of the spouse who is in resonance. Suddenly, instead of two communicating adults, there are two powerless children oppressed by a parental figure. In these moments, *choosing* to remain an adult in the here-and-now, *even if* we do not know how, is a very beneficial attitude for the emotional health of the couple. It is more beneficial than criticism. In fact, we want our partners to stop falling back into their childhood wounds so as to avoid constantly entering into resonance with them.

We are all bound by the frequencies of our genetic code. In my opinion, if our family has such an impact on us, it is undoubtedly due to the existence of an electromagnetic connection like the one that exists between spouses. I have noticed that the death of a parent often triggers a feeling of liberation in my clients. It is as though the deceased parents have left with their frequency rates and my clients find themselves freed from the influence of this frequency. They can decide to live outside of the frequency that was associated with the deceased person. They can choose to vibrate at a frequency unknown to the parents who have left them. After a parent's departure, they can continue to resemble this parent or choose to expand their perspective.

In general, the parent who dies leaves with personal, mental and emotional patterns. These patterns then dissolve and we can either reinforce them or we can stop and take the time to choose to manifest who we are outside of this influence. We can allow new forms to emerge that will correspond more closely to our own essence. In the following chapters, we will learn how to identify ourselves with our essence. We will finally be able to show ourselves adorned in our intrinsic frequency rather than in the numerous wounds we have suffered.

Keeping the Magnetic Charge of an Emotion

When the emotional body feels judged, it becomes paralyzed. It is possible to use the magnetic charge of emotions as the driving force of our intentions at zero point. I remember capitalizing on a moment of great sadness to command health. I went to bed and reserved a good hour of solitude for myself to allow my sadness to move without judging it or expressing it externally. By breathing and accepting it, I let the energy from this emotion flow without blocking it. Even if I was afraid of the strength of this sadness, I opened myself. I felt safe because, instead of experiencing it in public through mirror incidents that would have reflected it back to me, I allowed it to evolve in a controlled environment. Then, when I felt that I had reached the core of this sadness' emotional movement, I used its power to command health because I felt that I was coming down with the flu. I immediately experienced a sudden high temperature and then my symptoms faded.

If the emotional body is overloaded with stagnant emotions, the body will try to stimulate movement by causing an inflammation or pain. The block will inevitably be passed along to the physical body. We must stay with our emotions like a midwife next to a patient in labor. She patiently stays by her side, knowing that this hard work will be crowned by a birth. Emotional crises have the power to transform our lives. They can be marvelous transitions. There is nothing to stop us using the force of our past deceptions or our regrets to advance toward our goal. We can choose to forgive while feeling bitterness and anger. We do not need to change; we can use a feeling that seems pathetic to create a specific intention.

The Here-and-Now and Previous Reactions

With the benefit of a little hindsight, we can begin to see that our conflicts are remnants from our childhood. Sometimes we become hyperactive, our tone changes or we become impatient when we find ourselves feeling the familiar childhood powerlessness. In these moments, it is useful to ask ourselves to what age we have regressed. I had noticed that each time my father came to visit, I fell into a state that was entirely different from my normal state. One day, *I chose* to remain an adult during our encounters, *even if* I did not how. The first time I tried this, I constantly verified, deep down inside myself, how old I was. I ended up being able to be in the presence of my father without falling back into a childish state. Previously we had both been reproducing behaviors that dated from the period of my childhood. I finally succeeded in putting an end to the scenario in which my father embodied the disapproving parent, and I, the not so very self-confident child. We changed our dance and struck up an adult-to-adult relationship to learn to get to know each other in the present.

I have noticed that when my clients put up resistance at work, they are generally in **child** mode. It goes without saying that we have to be an adult to hold a job, even if we sometimes feel very small inside. In such cases, it is a question of an **unconscious order** from our default programs. A past situation has been transposed onto our current reality, which is then distorted. We enter a type of trance that brings us back to when we were five, ten or twelve years old, for example, regardless of whether we are now thirty or fifty, or even older. Our reactions and feelings belong to the past and have nothing to do with our current reality. I have observed that when I feel bad, my discomfort is often related to a situation that has nothing to do with the present circumstances. It may be a charge that goes back to my childhood or a future project that seems to be slipping through my fingers.

The first time I became aware of the phenomenon of the **unconscious childhood command,** I expressed the wish to be able to recognize programs from the past that resurfaced without my knowledge, so that I might be able to put an end to them. As a matter of fact, many signs indicate that I am not in the reality of the moment but rather in my childhood. With time, I have come to recognize these signs: feelings of restriction and powerlessness, my tone, my posture, the depth of my breathing, my doubts, etc. All these elements made me understand the need to implement a process that puts an end to these default commands that cut us off from the here-and-now.

Our default programs can also propel us into the future. For example, spring is barely over and I am already worried about whether my fall projects will come to anything. The lupines have barely poked out of the ground and I am already afraid of experiencing failure in September! But what exactly is it that always draws us into the future and prevents us from enjoying the here-and-now? An overload due to a default program powerful enough to create dissociation and have us lose contact with our body (something we all learned in our childhood). This default program can be of an emotional, mental, physical or spiritual nature, but the important thing is that by losing contact with our body, we become weaker because we disconnect ourselves from our vital energy, which then stops moving freely.

Take Maryssa, for example. She works in a university library. Her superior was so authoritarian that Maryssa had to file a grievance with the union. She consulted me because she wanted me to help her in her meeting with the union and her superior. In this case, we first had to determine Maryssa's emotional age in this conflict, that is to say, the age at which she had experienced a similar conflict. I therefore used kinesiology to find her age: three years old. Together we identified the person from her past who today corresponded to her superior. It was Maryssa's grandmother, who had lived with them. She had had an overbearing nature and liked to provoke. Maryssa's professional situation automatically brings her back to a period when she was too young to be able to defend herself. When we become aware of certain acquired attitudes, we can voice very precise commands, such as: "I choose to be present and safe in stressful situations while integrating the mental and verbal violence I experienced when I was young."

To free ourselves from our default programs and their tyranny, we have to make conscious choices. We must therefore work with conscious, rather than default, intentions. I invite you to rethink and review your default commands from a vibrational perspective. We can ask ourselves if our commands are truly in harmony with the vibrational rates of our superstrings, which emit the frequency of who we are. Do they vibrate in resonance with the frequency of our essence? From here on, we are starting a new phase in which we can attract realities that are in line with the quantum network that vibrates in harmony with the rate of our essence.

Chapter 5
Coexistence

When I teach the concept of zero point, I am always surprised to see that my students have a polarized conception of zero point, as though the negative magnetic charge of an emotion only existed to polarize the positive charge of an intention. Of course, it is useful in that case. But it is even more dynamic in the context of coexistence.

Coexistence is a fascinating concept, by virtue of which everything has the right to exist simultaneously, as in the case of zero point. During an expansion process, it would be a good idea to allow our slower parts to coexist with the quicker ones. We tend to exclude the former for fear that they will slow us down. In reality, if we integrate them into the process, they will balance the energy of expansion at zero point. Otherwise, we end up endlessly repeating the Chinese proverb that states that what goes up must come down. On the contrary, thanks to this new paradigm, our expansion can be continuous and circular, like the perfect model of the spiral.

We then find ourselves in acceptance mode. We accept everything we are and we accept all aspects of our personality. We no longer seek to get rid of a particular character trait or a painful wound. We accept all that we are, everything that forms the fabric of our being. We perhaps fear that, if we include our **darkness**, we will make negative choices. We have to remember that coexistence is the step that follows suit to zero point. We do not want to be polarized on one side or the other.

Coexistence of the Different Levels of Expansion

Zero point includes both expansion and stagnation. We work at zero point to voice commands, create new patterns or put an end to recurrent scenarios. Working at zero point provides us with assurance that our commands are effective, and from this certainty, comes an ever-growing feeling of well-being.

I noticed that after re-appropriating their power to command, my clients often experienced an expansion process at a professional or personal level. However, in this process, they would go through a period of success that was often followed by a period of regression. For example, one person had obtained a promotion, but had a tendency to polarize euphoria by setting aside her less ecstatic feelings. Disconnecting the negative aspect from a happy event that triggers an expansion hinders the expansion process. When the negative pole of an emotion has not been integrated into the feeling of expansion, it slows it down. It is as though the boisterous child you neglected to invite to your party were voicing his protest by making an infernal racket. This child will spoil your party if you do not deal with him.

According to the quantum model, every positive choice coexists with multiple possibilities, some of which are diametrically opposite to the choice in question. When I choose to trust my body, I must have this choice coexist with my former poor health, my fear of illness, a latent cancer, my blooming health, my phobias, my successes, my healings and my failures.

Expansion at Zero Point

Thanks to the Hubble space telescope, astrophysicists have proven that the universe is well and truly in expansion. We humans are subject to the same laws and we too are also going through a phase of expansion. Since the quantum observer influences matter, we too can influence our lives. I remain convinced that through the power of choice alone, human nature can access multiple possibilities for expansion that vibrate at various frequencies and exist in the latent state in parallel realities.

Expansion, however, is governed by cosmic laws. The more our projects expand, the more we should expect to feel the effects of this expansion at all levels, both dark and light. In fact, according to physicists, dark energy is a repulsive energy, which explains the acceleration of the expansion of the universe. This dark energy plays a crucial role in the process of expansion. It influences all aspects of our being. This means, therefore, that it will also accentuate our dark sides. However, it is easy to have all parts of ourselves coexist through the power of intention at zero point.

Chapter 6
Achievement and Self-love

Self-love at Zero Point

Some years ago, when I started to think about ascension (see Part IV), I understood that if they were to ascend, all the cells in our body needed to vibrate at the frequency rate of self-love. To achieve this, there is no need to become heroes. We can command, choose or decide to love ourselves the way we are, by accepting all aspects of our beings that we do not like. Of course, we can look in the mirror and criticize those aspects that appear less than perfect, but we can also do the following exercise. Take the time to examine and observe those aspects of your body that you do not like. Each time your eyes come to rest on a part of you that you do not like, state the intention to love this part *even if* you find it ugly, or... In a few weeks, you will find that you have a different relationship with your body. Magazines overflow with suggestions regarding how to strengthen our self-esteem. Among other things, we have to criticize ourselves less and work at reinforcing our self-image. Self-love at zero point is a broader concept that requires that we learn to consciously identify our good and bad sides while having them coexist.

There are some people who take their evolution very seriously and demand so much of themselves that I sometimes wonder in which direction they are really going. I myself tend to be self-critical. We all carry around inside ourselves an inner blame transmitted from generation to generation. Sometimes I wonder if we are progressing towards perfection or towards self-love. It is not by criticizing ourselves that we will achieve self-love, but rather by accepting each and every part of our being and by being gentle with ourselves through all our experiences. Self-criticism is a very polarized experience. We think we have to be perfect to be loved because we have not experienced unconditional love. How can we look at others without judging when we are so demanding of ourselves? Comparison is the most obvious symptom of the desire to accomplish amazing feats.

Some of my clients despair because they have not "accomplished their mission." They live in fact under the illusion that they have to accomplish great feats. The term *mission* blocks their creativity, intuition and freedom because it includes the implicit idea of an obligation. On many occasions I have had to treat clients to modify this demanding and oppressive code.

Our nature can be very dynamic without us having to accomplish a momentous achievement. Do we need to know whether our actions will save the planet? It is perhaps more useful to know whether or not they reflect who we are, in all our positive and negative aspects. Even if we doubt ourselves, it is preferable to act with a clear intention by adhering to the frequency rate of our essence. We will then define ourselves according to our frequency rate and our capacity to maintain this frequency, rather than according to the relative length of our list of achievements.

When we trust our essence, we become increasingly free to be ourselves, which gives us a feeling of personal power. We live in a society that is more and more regulated by the State and multinationals. This multitude of rules and laws generates a feeling of collective powerlessness that gives us the impression that we are losers from the outset. But we do not have to feel powerless. We can use the strong magnetic charge of this feeling of powerlessness to effectively command what we want according to our individual moral standards and in this way move on to another stage marked by greater personal power.

We alone can take this next step. It cannot be controlled from the outside because it takes into account our individual frequency rate rather than energy imposed by the great powers of this world. Our intention and our will to maintain a frequency rate that we have chosen, in keeping with our essence, will help us figure out what we have to do and how to do it without being constantly polarized by the judgment of others, self-criticism and the fear of laws.

Achievement and Self-love

In moments of financial insecurity, doubt or hardship, we sometimes lose faith in our essence. By encoding in our DNA the magnetic charge of the feelings of powerlessness and shame caused by these events, we will reach a higher level of consciousness. We will channel this energy by voicing the intention to have our essence vibrate even more strongly within ourselves. These difficult moments may well increase our faith in ourselves. It is not a question of religious faith, but rather of faith in our essence.

To work on ourselves, we need to have faith in ourselves. Of course, we can aim for a certain well-being, but we must not lose sight of the fact that our essence is attached to our being and is separate from our wounds and our environment. When we focus excessively on our wounds, we risk losing sight of this. Of course, we need to decode our wounds and understand their effect on our beliefs and reactions. But we cannot become discouraged or fall into the trap of feeling powerless by forgetting the pure essence of healing because it has become too technical.

There is a story behind each healing. Each one is a pilgrimage. Too often we think there is a magical formula requiring only one or two bits of information, whereas the pilgrim's path is strewn with ordeals. Each trial has its own lesson or leads to deeper self-knowledge. Emotional or mental realizations have but a single goal: to convince us of our divine nature. Pilgrims know that all the obstacles they meet along the way lead them towards this goal. There are no good or bad results.

It was through spirituality that I began to get to know myself better. At the time, I longed to merge with the vital energy that made electrons revolve around an atom's nucleus and that kept us alive. Each experience or obstacle reinforced my enthusiasm and my determination. It was easy for me to persevere because I believed in my divine nature. Everything that happened to me led me in this direction. Little by little, the decoding of my mental and emotional processes took over. This was necessary and natural. Yoga exercises, meditation and vegetarianism were not able to soothe every wound. My only regret is that I can see that most of us pilgrims have continued to identify with this stage. We are so obsessed with the desire to improve ourselves that we have lost faith in our intrinsic nature. Our journey is no longer mysterious and moving. Ecstasy has been replaced by an obsession with therapeutic achievement and results. Finally, mysticism has been downgraded by the fear of sects or by a need for success imposed by the corporate world.

It seems to me that we spend more time with our wounds than with our essence. Nowadays, after having exchanged our names and occupations, we introduce ourselves by recounting our mental suffering and our success in healing. We define ourselves in relation to emotional and mental codes. It is as though a single perspective dominates our internal landscape. When we move through this landscape, it is important that we stay properly centered inside ourselves rather than in our wounds. We are so identified with these wounds that we feel guilty allowing ourselves a few moments of rest and leisure that have no therapeutic goal. Internally, we think we must always improve upon ourselves.

What would happen if we learned to know ourselves, not to avoid an emotional, psychological, energetic or spiritual collapse, but to loudly and strongly emit the frequency rate of our unique and divine essence? We would certainly learn to live with dignity and individuality. We would stop aiming for perfection and would instead radiate our personal energy, with its unique and precious fragrance. We would give off a light that is different from the light of others, without comparing ourselves.

Having faith in our essence is a blessing. People who believe in themselves feel blessed. They are proud of themselves instead of worrying about their position in the therapeutic race toward results and achievement. In reality, to want to accomplish feats where healing is concerned is to abandon ourselves. Believing in our essence is not a luxury, since one of the fundamental laws of the Source implies that each of its elements vibrates at a unique frequency rate within the universal frequency.

To truly understand everything we are, we have to trust our values and have faith in our essence and in our own energy. Understanding our wounds and wanting to improve is all very well and fine, but the next step requires that we have faith in our essence and in our vital energy. When we start believing in our essence, we are automatically connected to the Source. Being convinced of the profound worth of our essence does not stem from the ego, but from knowing our personal frequency, the very frequency that keeps us alive and carries our signature throughout the cosmos. We can draw from our frequency the energy required to create our lives and have it coexist with the magnetic charge of our suffering.

Self-love at zero point is the ultimate recipe for having all our lives coexist, for returning to the Central Soul, for commanding with assurance, for making new choices and so on.

Our main objective is to maintain the frequency rate of our essence.

This frequency is unique, regardless of what we want to accomplish, regardless of our failures or our successes!

CHAPTER 7
CONSCIOUS ABUNDANCE

Conscious simplicity has received much media coverage in recent years. It has, however, much less impact than conscious abundance. What exactly is **conscious abundance**? It is a type of abundance based on our ability to take care of our needs and of ourselves. It is not about financial worth or how many things we own, but rather it is a question of magnetic abundance, by virtue of which our personal needs are fulfilled quickly and in such a way that our lives are balanced. It is a lifestyle rooted in self-love and dignity.

Self-love and dignity are not directly related with simplicity or how we spend money; rather, they reflect our capacity to manifest what we need to lead a balanced life. At zero point, when our objective is to change the way we take care of ourselves, we undeniably choose magnetic behaviors that bring about greater movement in our life. Changes take place more quickly and our commands produce results at greater speed, which in turn generate abundance through the law of entrainment.

When we manifest what we want based on our essence, we no longer dread the risks inherent to abundance and expansion. Our well-being carries our unique signature. Instead of comparing our lives and our successes with those of others, we become deeply satisfied. We identify with our feeling of fulfillment because it vibrates at the same frequency rate as our essence. Our prosperity expresses who we are because it is in keeping with our essence. We no longer fear losing ourselves in extremes.

A good many of the people around me no longer dare elaborate far-reaching projects for fear of exhaustion. However, if our manifestations are adapted to our frequency, we can embrace our successes without fear and without feeling overwhelmed.

Magnetic Abundance

We often hear about periods of expansion that were followed by periods of restriction. We have known people who have lost everything after having gone through a period of abundance. There are plenty of stories of lottery winners who, several years later, declare that they were much happier before becoming rich. When I watch the rise of a celebrity, I am always curious to know whether this rise is magnetic or doomed to failure. We only have to read articles about Michael Jackson to understand this phenomenon, or quite simply the story of Romeo and Juliet, whom death brutally tore away from great happiness!

Magnetic abundance is not simply a question of having money; it also entails the movement of a prosperous energy. The expression "to hit the jackpot" is a perfect example of polarized abundance, which reflects an energetic movement outside of the normal current of life. In magnetic abundance, everything coexists. The high points coexist with the low points, and both can be a source of abundance in their own way, just like the abundance produced by a large income.

We tend to divide our lives into pieces. We can imagine that our life is a big pie. One piece of this pie corresponds to moments of success while another, to moments of fatigue and less productivity. At these low moments, we tend to only see this piece of the pie. The piece that represents success no longer exists. However, by virtue of coexistence, fatigue is only one piece of the abundance pie, and this piece can coexist with another piece, such as success.

Conscious abundance has nothing to do with a region of the world or a professional or social status, but it does have to do with our childhood and our choices. From now on, conscious abundance must become an integral part of our personal moral standards. We have to decide, for ourselves and for others, that we have the right to meet our needs quickly and effectively. Instead of proclaiming that, from now on, we will take care of ourselves–a polarized state–we will cultivate an attitude and a way of thinking based on the conviction that we all have the right to abundance.

Thousands of people think that doing work they love is a luxury. This attitude blows me away. It is as though we were conditioned to working like slaves and accepting our reality without breathing a word. We constantly justify our powerlessness instead of using the energy generated by this feeling to command a job that suits us. Because our parents might have worked relentlessly all their lives, we do not see that it is possible to reject our slave conditioning and have a pleasant job. From now on it is possible to choose a new paradigm and to live in the present by having our belief in slavery coexist with abundance.

At the end of the course in naturopathy that I took in Montréal, the professor of ethics warned us that we should not hope to earn much of a living in our profession. I told myself, "A good thing I also studied naturopathy with a Californian visionary." Indeed, the latter claimed that if the therapists were not healthy or happy, the therapy would be mediocre. He was giving us the right to respect our own needs just as we would respect those of our clients. Thanks to this Californian teacher, I never believed that being a naturopath, rather than an executive in a prestigious firm, for example, would prevent me from sending my children to good schools or dressing and feeding them properly. Our limited abundance is sometimes due to a course or a social group that imposed upon us a perspective centered on scarcity. This perspective prevents us from using our free will to fulfill our aspirations.

A friend of mine who had just moved was asking himself, quite anxiously, whether he would be able to meet his mortgage payments over the years to come. I told him that he didn't need to worry because this new house was the concrete manifestation of his previous intention and it included his fears. Three years later, he and his wife received a large salary increase, which enabled them to do much traveling. What is more, they were able to renovate their home. These are signs of magnetic abundance where, step by step, intention by intention, the couple built their success by embracing their worries and fears as they arose. The most difficult part is to understand that financial insecurity, the fear of illness or weakness in general, are part of the prosperity wheel. It may be easier to visualize coexistence as a wheel or a pie cut into sections than as quantum realities in parallel universes, but the concept remains the same.

True abundance is that which allows us to be in contact with our inner voice in our more ordinary moments and not just during the significant events of our lives. Being in contact with ourselves and following the diktats of our inner voice is easily worth all material riches. I recently heard a singer being interviewed on the radio and she was explaining that she always trusted her feelings before making an important professional decision. She admitted that she did not always make wise decisions but that she systematically trusted her inner voice and was willing to accept the consequences of her choices, whether good or bad. If we constantly aim for perfection, we risk not hearing our inner voice when the time comes to make a decision. It is better to make decisions that are in keeping with that little voice than to be obsessed with a drive for perfection polarized by the desire for success or the fear of failure.

Vibrational Abundance

There are projects we feel so strongly about that we throw ourselves headlong into them without taking the time to check if they are in resonance with our frequency rate. We are so excited that we disregard our doubts. Take for example one of my clients, whom we will call Lorraine, who wanted to buy a piece of property within the context of an ecological project. She found the ecological concepts advocated by the developer appealing, but she did not take the time to feel whether this project was in harmony with her own frequency. We are sometimes so taken in mentally by an idea that we disregard all its other aspects. We refuse to listen to our hesitations for fear that they will prevent us from reaching our goal. I helped Lorraine leave aside the form of the project and its underlying concept to be able to feel this project on a vibrational level. In letting herself be convinced of the ecological objective of the project, Lorraine had yielded her power to the property developer instead of remaining centered and commanding safety at the vibrational level. She therefore *voiced the intention* to live in a home that would be in resonance with her essence and that of her family *even if* she didn't know how to do this.

By choosing projects that are aligned with our frequency, we give ourselves permission to express the totality of who we are instead of leaving this decision to our inner child and her/his unconscious programs, while knowing that she/he does not have the skills required to make a choice that suits our essence. We should therefore consciously voice the intention that our future choices be aligned on the frequency rate of our essence; in doing so we will feel safe on a vibrational level. We must expect, however, that doors will irrevocably close when a situation is not aligned, and that they will open without effort when we manifest, through conscious intentions, situations that correspond perfectly to our frequency.

By daring to command that our projects be aligned on the frequency rate of our essence and that they be vibrationally safe, we access a new form of continuous manifestation. Our projects are no longer in line with the vibrational reality of someone else, such as a convincing salesperson, but with our own essence. Our commands at zero point produce impressive results. Our undertakings do not come to a sudden halt after a certain time, but continue to adapt to various frequency modulations. We are not constantly initiating new projects because the former ones are no longer in keeping with who we have become.

The need to feel that our undertakings resonate with our essence and bring us vibrational safety is not a false desire. It is not based on materialistic constraints or a feeling of insecurity. It is an essential choice!

CHAPITRE 8
SELF-SUPPORT

TRANSITIONS

The *American Heritage Dictionary* online defines the expression "rite of passage" as "*A ritual or ceremony signifying an event in a person's life indicative of a transition from one stage to another, as from adolescence to adulthood.*" When we enter a new stage of our lives, whether professionally or socially, we have to give ourselves the support we need because our society has given up rites of passage. We hold the unconscious belief that our lives are run on automatic pilot. Therefore, we passively submit to periods of turbulence and transition. When I honor a transition period, I give myself time to go through it and I treat myself with that much more compassion.

To give ourselves self-support, we must take the time to recognize the importance of the transition we are going through, because this transition will lead us to a new phase of our development. What we are experiencing is sacred as long as we decide that this is the case. Similarly, this transition will be a commonplace occurrence if we make it so. From the moment we recognize a transition as a sacred event, we commit to granting it patience and attention. Even when we wake up with an anxiety attack in the middle of the night, even when we are tormented by our worst fears, we can state our intention to have our well-being coexist with these states of mind. During troubled times I tell my vulnerable self, "I am here for you; everyone could turn against you, but I am staying right here. I am here for you! I do not fear your pain!" I tell the part of myself that is not so brave, or that is less skilled, that it can count on my support to get through this transition!

Many of my clients schedule appointments with me on their birthday. This anniversary is universally recognized as a transition. They have a feeling that they are starting something new, but they do not always know how to stay in touch with themselves during this process. Yet, the more they stay in touch with themselves, the easier it will be. To provide self-support during a transition period, we have to ensure that all parts of ourselves that are going through this transition have the opportunity to coexist. The more we evolve, the more we become attentive to ourselves as we go through transition periods. You can tell good therapists by the way they take care of themselves throughout their growth processes.

Self-compassion

To reclaim our emotions, we need to be present to their negative charge. Otherwise, the void thus created may be filled by negative charges from the collective consciousness, or be exposed to default programming. Instead of leaving the charge wide open, it is better to integrate it in our choices. For example, suppose that one morning we wake up tormented by doubts. We can choose to have our health coexist *with* our inner doubts instead of being depressed for the rest of the day.

Spending time with the vulnerable part of ourselves, which contains our sense of powerlessness, is not a full-time occupation and does not mean that our current reality will be permeated with a continuous sense of helplessness. It simply means that we will be in contact with all aspects of ourselves and that we will need to work from the core of our being to define the vision of our central intention. If I am not in contact with my vulnerable side, it will take control using default programs. If I *am* in contact with it, however, I will be the one who takes command, entering into a dialogue with it, thus redefining my reality.

For example, I can tell this vulnerable part of myself, "I will spend time with you; I will stay with you," and in the same way, "*I choose* to express my power *while* staying with the part of myself that is suffering at the moment." The simple fact of stating an intention that includes my vulnerable side (the part of myself that suffers) enables me to avoid default programs and to affirm my power. In other words, rather than letting my vulnerability limit my potential, I use it to manifest my intention and it then becomes an asset in my life.

By doing so, we ensure that there are no empty spaces through which a foreign negative charge could settle. When we stay present to our own emotions and work with them, we can experience true compassion. It is very difficult to stay centered when a part of ourselves is completely open. According to the quantum model, in which all realities exist simultaneously, all probabilities must coexist in order for nature to be in harmony. To be centered, balanced and successful, all parts of ourselves have to coexist.

We could serve others better if we could be present to ourselves. In our moments of vulnerability, the more we are available for ourselves, the more we can feel compassion for others. When we assume the role of the Good Samaritan, we use the suffering of the other person to become present to ourselves and to all that this suffering evokes in us. On the other hand, through our compassion, others can get in touch with themselves. If we are comfortable with our own unease, our friends will be able to cry and we will not require them to stop. We will be able to accompany them through their period of transition without interfering in the process.

One day, a friend of the family who was about to go on her first trip abroad came to our house for dinner. During the meal, she felt strong emotions well up inside her. She burst into tears and expressed her fear that she would arrive in South America and feel lost, as though on an island in the middle of the ocean. Our friend allowed herself to express the whole range of emotions that emerged because we were not polarized by emotional censorship. We told her to cry, "It's the first time you're going on a plane and you're leaving for several weeks. You have the right to be afraid. Better express your fear now and make it coexist with your joy than ignore it." Soon after, our friend left. General strikes broke out in the country she was staying in and several neighborhoods were evacuated. She always felt safe, however, and was not affected by these events.

We have to understand that when we experienced childhood traumas, there was no one to help us work out our feelings. The adults around us could not confirm the legitimacy of our feelings or did not explain that our grandmother had her own problems to resolve. We were left alone with our suffering. If we had received support at the time, we would not be endlessly grappling with the same conflicts as adults. If as a child for example, I experienced a traumatic situation of which I am no longer consciously aware, you can be sure that life, which works in a spiral, will present me with this situation again and again in a different context. It is a bit like playing Clue: we have to discover who did what, in which room and with which weapon.

It is in moments like these that we have to **support ourselves**. I suggest to my clients that they take a pillow, a cushion or a teddy bear and withdraw into a calm place. I recommend that they take the teddy bear in their arms and say, "Your grandmother (your father, your mother, etc.) terrorized you when you were a child and was not there for you. I am here now. I have come back for you. I know that you are suffering. I am here for you, dear, and I do not fear your pain. You can cry until you can cry no more; if you felt abused, I believe you. I am here for you."

When we behave as adults, it is possible to allow our inner children to cry until they can cry no more and finally gain access to their repressed feelings, **without censuring them**. Next, we promise these inner children to return every day, for as many days as needed, and rock them and stay with them for five to ten minutes at a time–not the whole day. Our adult selves must also ask these children not to interfere in our current adult lives. From now on, when we go to work, for example, we can choose to be in the here-and-now *while* carrying the weight of the criticism we were subjected to as children. Having taken this decision, we can return to work and voice the intention of being respected *while* accepting the insecurity we felt as children. Little by little, even if our professional situation does not improve, we will feel less and less like victims and we can go to work feeling happy.

This type of self-support has been of great help in my life. It has allowed me to become my own best friend, my strongest support. In difficult moments, I repeat the following like a mantra: "I am here for you! There's an adult who is here for you. I do not fear your pain. I am here for you. I am with you." This is a form of meditation that promotes self-support. We were not always alone when we were young and the adults were not all bad all the time. They were often overwhelmed and simply unable to be aware of what we were going through. This is why we need to go back and assume the role of the attentive parent. During our childhood, we experienced traumas on our own; we were too small to have any idea of how to resolve them.

When we go back to unravel our childhood, we have to be able to state to our inner children that they are not alone and that they finally have an adult by their side. If I sometimes wake up at night and feel fear or any other emotion, I automatically adopt this self-support mode. I repeat my *mantra* until my emotions subside. I end up falling back asleep with a feeling of well-being because I have experienced a supporting presence.

We cannot erase our childhood traumas, but we can support our inner child with our presence and compassion. We can also keep the magnetic charge of our traumas and use this charge to magnetize new scenarios in our adult life. I can be creative and experience new events without feeling afraid of being isolated if something goes wrong. I know that I will be present to myself, that I will provide this presence with compassion, and feel no shame for my emotional reactions, even if it feels like I am not making any progress or if I hesitate to accompany myself for a certain time. Self-love and self-acceptance, when balanced by a lack of self-love and self-acceptance, prevent me from adopting a superhero attitude and instead allow me to remain present to myself with compassion. I will not wait to be perfect to choose clarity because I would most probably be waiting a long time! Since I have adopted this behavior towards myself, I feel calmer. I have an increasing sense of authenticity and I appreciate it in people who are not afraid to feel discomfort in the presence of others because they can remain present to themselves instead of expecting to be taken care of.

This way of thinking does not apply only to women. Many men who consult me have suffered from not being able to satisfy the demands of their mothers. As adults, they feel inadequate or defensive before the conscious or unconscious demands of their spouses. They need to return to their childhood traumas, spend time with their inner child and tell him: "Poor you. Your father was never there; your mother was overwhelmed. I am here for you now. It must have been so hard trying to be the boy your mother wanted you to be. You must have doubted yourself. The adult that I am knows that it is impossible for a one-year-old child to meet all these expectations."

I have witnessed the gradual evolution at zero point of individuals who did not know how to delegate and were suppressing their frustrations. They became more spontaneous. Their inner child, less isolated, was no longer always in the foreground and they gradually were able to develop warm, adult relationships, based on mutual cooperation and authenticity.

Much illness and suffering could be relieved by self-support. One of my clients, who had suffered from cancer and who had healed herself using alternative methods, did not outlive the shame that engulfed her when she had a relapse several years later. She could not manage to remain present to herself and did not accept that her childhood wounds still require so much attention. These wounds are like a boisterous child: if we don't deal with him, he shouts even louder. If we accept to be disappointed without being ashamed, we will discover what lies hidden behind our perfectionist attitude.

In this type of decoding, we not only need to take into account the significant moments of our childhood, but also other types of memories, such as transgenerational memories. Having learned to support myself in regards to the painful periods of my childhood, I was able to continue this process with my transgenerational, social and historical memories. Of course, we need to decode and interpret our wounds, but being self-supportive is the most important task that subsequently falls to us.

When I have a nightmare, instead of feeling relief once morning has come, I welcome my nightmare. I stay present to my fear because I know that the nightmare sends me back to my old wounds. If I linger over them, rather than sedate myself, and welcome the magnetic charge of the nightmare, I take advantage of this energy and thus ease the after-effects of my past traumas.

We tend to hope that our problems will magically disappear. This attitude, in addition to not being very constructive, also leaves us open to experiencing the return of our problems like a failure. Our childhood wounds will not disappear. They are engraved in our memory. I sometimes feel that my wounds are reactivated, but it is occurring less and less often. In addition, the shock is less intense and the effects do not last as long. I am no longer afraid to see a wound resurface because I am my own parachute, now that I have chosen to support myself.

As soon as we accept a particular aspect of our human experience, whether it be fear, boredom or anger, and decide to have this emotion coexist with the rest of our life rather than get rid of it, not only does this aspect no longer cause a problem, but we no longer seek to get rid of it as fast as possible. When an incident occurs that triggers this emotion, we welcome it and our life remains fluid. The quantum model speaks of matter that is in movement and that constantly transforms itself under one influence or another. Our life on Earth continually triggers a range of emotions. Why not be present to these emotions and experience them fully?

The human challenge in the face of any suffering consists of finding new ways to deal with it. There is a powerful solution: do not wait to be **saved**, but become conscious and responsible. By responsible, I mean being able to **meet** our personal needs. And, especially, allowing ourselves to feel our emotions without responding by taking action right away. We can give ourselves time to try this new approach without being too harsh on ourselves. After all, the human experience is complex and we do not live in a perfect world.

We all carry an ideal image of what life should be. As soon as we see a couple that appears happy, we unconsciously imagine that there is a perfect relationship model. In fact, we would like to believe that there is another avenue other than supporting ourselves in moments of darkness. But the perfect couple does not exist; there are only authentic couples made up of two people who know how to support themselves individually and mutually.

I take pleasure in imagining a society of adults who would take the time to support themselves in their moments of distress so as to manifest a compassionate world. The more I support myself, the more I can support my children and guide them without interfering in their lives; the more I feel empathy and tolerance toward others, the more I want to take care of our planet. Love and self-support might very well be a perfect scenario for our life purpose on this planet.

Supporting Myself Rather than Changing the Other

When I was a young mother, I always thought that I should strive to improve my children. As I grew older, I understood that this was not so. Instead, I had to fix myself. I also had to be self-supportive. If people want to establish a friendly climate at work, they should choose to work in an atmosphere of mutual help and support, even if they do not know how, rather than ask those around them to be the ones to work on changing themselves. They will then have a greater chance of succeeding. If they think that they have a right to receive support and love, they will intentionally manifest a working environment made up of allies rather than adversaries.

Take Florence for example, whose children constantly quarreled with their father. After trying time and time again to bring them to communicate calmly, she turned the situation around. In the very first place, she asked herself, "Who am I when my children are in conflict with my spouse?" She felt irritated, disappointed, lost and, finally, she recognized that she thought she was a bad mother. On the one hand, she understood that the problem was between her husband and the children. It did not belong to her. On the other hand, she decided to work on herself because the situation triggered emotions in her. She therefore used her own anger as the driving force of her intention to manifest peace in her home. She used the energy of her frustration to resolve the conflict. Often, the problem is that we wait for the situation to change instead of using existing tensions as the driving force of transformation.

Certain questions can help us gain perspective and take care of ourselves during a conflict or during moments of tension: Who am I when the people around me quarrel? Who am I when my spouse quarrels with my child? Can I feel my legs? Can I feel my feet? Living in the moment without judging the situation allows us to remain present. If we have been hurt during our childhood, it is possible that we may have learned to dissociate ourselves from the situation instead of remaining present. As adults, we can choose to breathe deeply and remain present to ourselves.

Painful and joyful moments can begin to coexist within us. They both have their place and must not be overshadowed. We must not put an end to suffering, but rather to the avoidance or rejection of our suffering. Some think that they need to heal their old programs to avoid suffering but, in reality, supporting ourselves in our suffering is much more beneficial for our whole being. We must let suffering be and **use it** to make new choices. We do not want to let our vulnerability and our past suffering unconsciously dominate our current lives. In other words, our goal is to defuse our default programs. It is not to eliminate the vulnerability and suffering at the root of these programs.

One of my clients had problems with her son who had been prescribed hyperactivity drugs by doctors. Someone had told her that parents were almost always at the root of a child's problems. She therefore turned the situation around and began to focus on herself using intentions at zero point. She gradually transformed herself by having her fear that her son would turn out bad coexist with the conviction that he would have a wonderful future. Result: her son changed so much that the school principal asked her to give a conference on the means she had used to help him.

This is a very important point. To create effective intentions, we need to be able to stay tuned to our vulnerable side. This also implies that we welcome our vulnerability and allow it to remain present in our bodies rather than avoid it through mechanisms of denial. During difficult periods, we tend to dissociate ourselves from our bodies. We learned this denial mechanism during our childhood, the period when we were the least bonded to our physical selves and when it was easy to "leave the body." However, as adults, we cannot be open to our vulnerability if we are disconnected from our body.

An intimate relationship is a privileged space for self-support. Without this support, the relationship is affected by judgment, criticism and disapproval. It is very easy to feel criticized by our partners and to feel that they are not there for us. In fact, our partners behave toward us as we behave toward ourselves. This magnifying mirror is the ideal opportunity for taking ourselves in hand and loving ourselves. How can the other love us when we are so lacking in self-love? Intimacy is first of all an encounter with ourselves. We unconsciously seek intimacy so that we can be ever more present to ourselves. We are here on Earth to learn to love ourselves, know ourselves and support ourselves. The other person is our master in this art. This person can only do for us what we do for ourselves. The road to intimacy is not useless. Each step reveals the key to the return to self.

We would all have liked to have known a great love like that of Romeo and Juliet or Tristan and Isolde. Yet these great lovers died before even experiencing a relationship on a daily basis. You may have noticed that we know nothing about the life of Cinderella after her marriage to Prince Charming. Our only models for intimacy come from our parents and grandparents. They never thought about the mirror effect generated by their lives as a couple. We are the first generation to do so. Therefore, instead of reacting by saying, "How horrible; how heavy!" we could think, "What an extraordinary thing!" In this context we are obliged to choose the new paradigm of zero point, in which our expectations and disappointments coexist in our relationship to self-love. If our couple needs inspiration, we will command it instead of waiting for the other to hand it to us on a silver platter. Step by step, over the course of our errors, we will finally grasp how to take care of ourselves and love ourselves while being in a relationship with the other.

Life works like a spiral. From time to time, it brings back the same emotional patterns that have hurt us during our childhood. We are endlessly confronted by the same unproductive scenarios and the more we move forward, the more it becomes imperative to face them, to remain present, to support ourselves and to make different choices.

Take Claudia for example. Her spouse did not like her going out without him. In viewing things from the perspective of the spiral, Claudia understood that her father had not been free either. Her mother, a domineering woman (like Claudia's husband), had reproached him daily for the time he spent at work, with his colleagues, playing golf, etc. As the years went by, her father had taken less and less pleasure in being at home. Claudia understood that she was carrying her father's legacy; he had never managed to reconcile pleasure and freedom with intimacy.

When we are young, in our eagerness to become adults, we start acting like grown ups. When she understood this pattern, Claudia first took the time to release herself from the choices that she had learned from her father and subsequently chose to manifest a new model at zero point using her acquired behavior as one of the magnetic poles. By undoing the unconscious connections that tie us to the past, we can design a path for ourselves that has not necessarily been taken by the people in our childhood. To do so, we have to spend time with ourselves and understand what would bring us pleasure. To spend time with ourselves, we have to free ourselves from conflict with others, turn the mirror back onto ourselves and *choose* to accept ourselves *while* not knowing how to love ourselves.

When Claudia questioned her paternal model, I asked her whether it was possible for two adults to be free while being parents. She answered in the affirmative. We explored all sorts of ways to put an end to her old model. For example, when her husband would do something for himself, she was in the habit of wishing him a good time. Since he was incapable of reciprocating, she decided to wish herself a good time even if she was afraid of him, like her father had been afraid of her mother. One day, before going home, she told her inner child that she wanted to have a discussion alone with her husband regarding how she had spent her evening, and she asked her inner child to go into another imaginary room. She told her, "When I was small, I suffered so much from my parents' quarrels that I don't want you to stay there expecting the same situation to keep repeating itself forever." And then she stated, "I choose to be an adult when I go home this evening and *I choose* to remain in my adult state *even if* I don't know how."

Even the best couples, even the most intelligent people need to remain vigilant to avoid these recurring scenarios. A relationship teaches us how to take a step back and make new choices to avoid re-living our parents' scenarios. Today's worldwide confusion forces us to question our old relationship models in order to develop a better future for the Earth. This is the advantage of being conscious. Each time we put our beliefs and attitudes in order and choose more conscious behaviors at zero point, we help society and those around us by emitting a new conscious frequency.

Vibrational Communication

Communication is a perfect example of a vibrational frequency. Like an intention, it emits a frequency rate. But, since we all emit our own frequencies, this makes communication difficult. Nowadays, it is noticeable how hard it has become to receive proper service in restaurants and stores, but we can change this state of things thanks to our essence. When people emit the unique frequency rate of their essence, they give off a certain harmony that triggers a specific vibrational response. By being self-contained, these people do not create antagonistic reactions. Since they are in harmony with themselves, regardless of what they might be going through, they are better able to understand the positions of others. They can choose to remain in their own frequency while having it coexist with that of others. Thus, instead of criticizing the poor quality of service, we can use every frustration we experience to command at zero point that our personal frequency be amplified and thus choose the frequency at which we would like to communicate.

Certain means of communication help us stay in touch with ourselves. For some people, communicating is easy when they teach or work, or when they help someone, when they are alone or in the process of selling something, for example. In certain spheres of our lives, we are in greater resonance with ourselves and we communicate effectively. However, when we proceed to another form of communication, such as the one used in intimacy, the frequencies that were acquired during our formative years prevail over the frequency rate of our essence. These frequencies, acquired through contact with our environment, create a distortion and create an energy disturbance that aggravates the person with whom we are communicating. In these moments of turbulence, it is preferable to withdraw for a few seconds to consciously change our frequency rate and replace it with that of our essence by using an intention at zero point. This is an excellent way of recycling the energy from a communication-related conflict or a feeling of powerlessness instead of withdrawing into ourselves or making violent remarks!

I have asked spouses how they managed to talk to each other without hurting each other when they did not agree. They said that they aligned themselves on the same frequency. As a result, their diverging opinions did not disturb their frequency rate. They managed to understand each other at a fundamental vibrational level, which decreased their mental or emotional discord.

If we keep working at it, we will manage to adopt the new paradigm of vibrational communication. By choosing to have the frequency rate of our essence vibrate through our words, we will invent a new form of relationship based on mutual respect and appreciation. We will be heard and understood properly. A conscious intention combined with a frequency rate that reflects our essence is without a doubt the solution to communication problems. By choosing not only to change our way of communicating but also to emit at a precise frequency, we increase our chances of being understood and of understanding others. In choosing a vibrational mode of communication, we emit waves that are in harmony with those of the people we are in communication with. If we look for this type of vibrational exchange, we will know how to respond in an unbalanced situation either by interrupting the exchange or pursuing it.

There are excellent communication techniques that can teach us to communicate effectively. They can teach us how to better communicate with others, but since the frequency of what we communicate precedes our words, it is this specific vibration that needs to be modified if we want to become effective communicators. Through this approach, we will listen and speak with all our cells. We will listen to the vibration of communication. We will feel the vibration, and the words that come out of our mouths will have a conscious frequency rate.

I find that when, on the one hand, the antagonistic reaction of others forces me to modify my approach, it also gives me, on the other hand, the opportunity to choose the reality I wish to manifest. First, I change my approach with others, and then the next stage brings me back to myself. By continually turning the mirror back towards ourselves, we end up seeing ourselves. That is when true compassion is born. To free ourselves, there is nothing more powerful than thorny situations or family conflicts. In intimacy, we are confronted more with ourselves than we are with the other. It is our own deep anxieties that we perceive, much more than the other's wounds. It is in bearing our wounds–with vulnerability and not necessarily with courage–that we finally *choose* to love ourselves and remain present to ourselves *even if* we don't know how! Gradually, this care and presence will lead us to our essence.

Certificate of Conscious Living

I designed the following Certificate of Conscious Living for my clients who decided to support themselves by being present to their wounds to make new choices afterwards.

This diploma underlines their determination to know themselves better and, especially, to support themselves through their dark periods. I want to celebrate this achievement. We are all connected to one another and when people change their behavior, we all benefit. This is why we have to celebrate this event.

One of my clients who was suffering from breast cancer admitted to me that she knew the emotional cause of her illness, but that she was unable to support herself in her pain, which was too great. She finally chose to face it while supporting herself and that marked the beginning of her healing. I therefore offered her a Certificate of Conscious Living. I wanted to recognize her effort and to congratulate her because she had brought to light the repressed violence of her childhood. When she succeeded in supporting herself in her suffering, it was as though she had just uncorked a bottle of champagne. Gradually, she connected with her past and she eventually healed. I had warned her that she would need a good dose of compassion and patience toward herself. After this transition at zero point, her healing quickened because things were moving and my client had changed frequencies. These moments are unique; they need to be celebrated!

Certificate of Conscious Living

AWARDED TO ______________________

FOR HAVING ACCEPTED TO FEEL THE PAIN CAUSED BY A CHILDHOOD WOUND IN RELATION TO

____

____

WHILE REMAINING PRESENT TO HERSELF/HIMSELF.

SIGNED IN ____________________ ON ____________

BY ______________________________

PART II

THE PROCESS OF SELF-LOVE

CHAPTER 9
BOUNDARIES

SELF-LOVE

To experience self-love, we need to know ourselves. How can we love someone we don't know? The question is, how do we go about getting to know ourselves? I have come to understand that to do so, we have to first eliminate all that is not our own within our circle. By circle I mean our *bubble* or energy field. If there are too many people in that space, it is impossible to discover who we are. Whether in our professional lives or in our relationships, it is easier to start loving ourselves if we are not in a fusional relationship with someone else.

In fact, the ideal relationship between two people occurs when two individuals choose to be together but with the essential goal of self-love. The partners are close while remaining separate. They coexist but they maintain individual frequencies. The only prerequisite: an individual, unique frequency rate. This profound individual essence vibrates at a frequency that calls to us, calms us, heals us and connects us to the Whole while enabling us to coexist with others.

To experience self-love, we need boundaries. Without boundaries, we may tend to take care of others instead of being self-supportive. The process of self-love is a personal process that requires that we emit the frequency rate of our individual essence. This does not mean that we will no longer take care of others, but we will do so while vibrating our personal frequency rather than merging the frequency rate of our essence with that of another. This cross-linking of frequencies creates a third one that is no longer ours.

On a beautiful spring day, I was skiing alone while listening to music. While on the ski-lift, I would bask in the sun, and then I would ski down the mountain to the sound of wonderful music. I was in an ecstatic state, concentrating on my essence and my interstitial void in these extraordinary conditions. Suddenly, I came across two skiers whom I knew. They looked at me without seeing me. I wondered if I should greet them or not. I was afraid of spoiling my state of bliss by speaking with them. I heard them chatting in the chairlift in front of me. I then consciously decided to ski with them on the condition that I would be able to keep my euphoric state. When we arrived at the top, they took off without seeing me. I had *voiced the intention*, if I skied with them, to have my frequency coexist with theirs *even if* I did not know how. I had given this experience priority because I was not yet sure enough of myself to combine my frequency with theirs. I knew that I could have a good time with them, but my state would nonetheless be changed. This is why I saw them leave without regret. We need healthy boundaries to maintain the frequency rate of our essence while in the presence of someone else's frequency.

We do not become unsociable because we establish boundaries. Rather, this is the perfect tool to use to choose and maintain a frequency rate that suits us. In fact, there is a good chance that our circle of friends will expand. Instead of refusing invitations for fear of feeling oppressed by some of the guests, we can use the power of intention and *choose* to close our circle *even if* we do not know how. We are not looking for self-love to isolate ourselves, but rather we are seeking to experience self-love as a tool that allows us to coexist with frequencies that are different from our own.

The Alchemical Law of Unity and Boundaries

The road to intimacy is filled with fusional traps. The quantum approach suggests a model that is closer to the alchemical model, which advocates Unity rather than fusion. By virtue of the concept of Unity, we are part of the Whole. We communicate with the Whole. We give and receive from the Whole. The Whole can speak to us, send us information, and vice versa. We are in contact with it and in it. We are one with the Whole, while keeping our individuality and our own vibration. As individualized frequencies, we are an expression of the Whole. This means that our energy has its own tonality, its own color, within the Whole. Every aspect of Creation has a vibrational frequency. We will continue to vibrate at an individual frequency rate even if we become one with the Whole. To ascend, we have to learn to identify completely with this frequency. To do so, we will first need to transpose this quantum model into our relationships, *even if* we do not know how.

Imagine that the person with whom we are in a relationship provides us with a feeling of freedom and excitement. By virtue of the principle of Unity, we can draw these qualities from the Whole through the power of intention without having to maintain a codependent relationship with this person. Without having to put an end to our relationships, we will become our own source of freedom and excitement and we will reach a new form of relationship, one that is free and inspiring.

Establishing Healthy Boundaries

The following exercise revealed itself to me when my seven-year-old son complained that he was depressed. This astonished me because at the time we were going through a rather serene period. I suddenly realized that this state did not belong to him but rather to his best friend. I created this approach so that I could teach him how to release his friend's emotions. The following day, I did the rope exercise with him and his father so that he could learn how to recognize signs of energetic interference in his personal energy field.

My son clearly felt the difference between the discomfort caused by open boundaries and the well-being associated with closed boundaries. He therefore had no difficulty choosing to close his boundaries again and to discard all that did not belong to him. Several months later, I noticed that he was making new friends more easily. I wondered how many children suffering from attention disorders or behavior or emotional problems simply had wide open boundaries. From then on, my son was able to take a healthy distance whenever his friend was not doing so well.

When we tried the exercise with the rest of the family and with our friends, the outcome was monumental. It allowed us to undo unconscious codependent behavior.

While doing this exercise, the participants feel such a difference depending on whether their circle (which is made up of ropes and corresponds to their bubble or energy field) is open or closed, that they can code the kinesthetic difference between being sealed or not, which they have sensed in their bodies. In addition, once they have identified what is obstructing their circle, they can consciously choose how they prefer to deal with it. I noticed that, in general, the energy of a person suffering from the flu or another contagious disease does not extend beyond that person's circle. It is possible to close our space and keep it healthy by voicing an intention to do so *even though* we don't know how.

A computer virus has no power as such. It must first have access to programs. This virus needs first of all to infiltrate the network, scramble the waves, and then infect the network. The same is true for us. For example, miasms, vectors left by a flu virus, parasites or someone else's emotional baggage can sometimes invade our network and scramble our frequency. After having identified what does not belong to us within our circle, we can get rid of it. Afterward, we will clearly see the emotions or feelings of discomfort that really belong to us. In acknowledging what is ours, we will then be able to use this magnetic charge to maximize the potential of our intentions and our commands.

Before getting down to work, remember that the goal of the rope exercise, in addition to stabilizing your boundaries, is first and foremost to lead you toward ascension. We need to establish healthy boundaries to be able to have the frequency rate of our essence vibrate within ourselves without being subjected to interference. Our aim is ascension rather than personal growth.

Exercice 1

ROPES (ESTABLISHING HEALTHY BOUNDARIES)

In this exercise, a couple or several people sit on the floor without touching one another. Each person is seated in his or her individual circle formed by a rope.

PART 1

CLOSED CIRCLE - OPEN CIRCLE

1. Ask the participants to close their circles by bringing together the two ends of their ropes.
2. Ask them to close their eyes and inhale deeply while observing how they feel physically and emotionally. Invite them to notice their breathing.
3. Tell the participants to open their circles and close their eyes once again while observing how they feel physically and emotionally. Invite them to notice their breathing and to observe the difference between the two states. *At this stage, most people feel discomfort–even young children.*
4. Ask them to close their circles once again when the situation becomes too uncomfortable.

At this stage, most people feel discomfort–even young children.

PART 2

CLOSED CIRCLE – OPEN CIRCLE: CLOSE TOGETHER

1. Invite the participants to close their circles by bringing together the two ends of their ropes.
2. Ask them to sit very close together. The circles can touch, but each person has to remain in her/his own closed circle.
3. When they are very close, tell them to close their eyes and to breathe deeply while observing what they feel physically and emotionally. Invite them to notice their breathing.

4. Next, ask them to open their circles while remaining very close to one another, and then to close their eyes again while observing what they feel physically and emotionally. Invite them to notice their breathing and observe the difference between the two states.
5. Ask them to close their circles once again as soon as the situation becomes too uncomfortable.

PART 3

CLOSED CIRCLE – OPEN CIRCLE: HOLDING HANDS

1. Invite the participants to back up about a meter while holding hands, then to close their circles by bringing together the two ends of their ropes.
2. Ask them to close their eyes and to breathe deeply observing how they feel physically and emotionally. Invite them to notice their breathing.
3. Next, tell the participants to open their circles while remaining close to one another, and then to close their eyes again while observing how they feel physically and emotionally. Invite them to notice their breathing and observe the difference between the two states.
4. Tell them to close their circles once again as soon as the situation becomes too uncomfortable and to hold hands once more.

PART 4

CLOSED CIRCLE – OPEN CIRCLE:
EXCHANGING A CURRENT OF UNCONDITIONAL LOVE

1. Ask the participants to back up about a meter while holding hands, then to close their circles by bringing together the two ends of their ropes.
2. Invite them to have a current of unconditional love move through their right hand and have it come back through their left hand. They place their right hand, palm down, in the left hand of their neighbor. The left hand will be held palm up.
3. Ask them to close their eyes and breathe deeply while observing how they feel physically and emotionally. Invite them to notice their breathing.

4. Next, tell them to open their circles and close their eyes once again while observing their capacity to emit and receive all this love with their circle open.
5. Ask them to observe how they feel physically and emotionally. Invite them to notice their breathing and to observe the difference between the two states.
6. Tell them to close their circles again as soon as the situation becomes too uncomfortable.

This exercise is a way to feel kinesthetically what intimate separation can be like. During a workshop, one of the students had a revelation. On the one hand, she felt deep compassion toward the other participants when she emitted a current of unconditional love while the ropes were closed. On the other hand, she was surprised to discover that this stream of love was interrupted as soon as she opened her circle!

In general, participants felt that when their circles were closed the energy was more fluid, the energy exchanges were better and they found it easier to perceive and be aware of the others. One mother, wondering how it was possible to be a proper caregiver with closed boundaries, understood that closing her circle in fact increased this capacity. Some people felt great weakness when the circle was open.

I invite you to try the open and closed rope exercise to have a kinesthetic experience. The kinesthetic sensation triggered by opening or closing your circle will be inscribed in your cells. You can go back to this inner feeling when the need arises to check whether your boundaries are open or closed.

Remember that we do not have to be heroes *of the closed circle*. We do not have to be perfect or mentally understand everything we have felt. It is possible that we will feel ambiguous or vulnerable, and that we will need a moment to integrate our experience after ending these exercises. We should give ourselves this moment of integration! It is not always easy to feel that we may need to close our circle and go within. Losing ourselves in other people's space can weaken us. It does not foster well-being. To bring ourselves back within, we will have to discover who we are and understand our personal needs. We can choose the context of zero point while knowing that the question of our identity can coexist very well alongside our doubts.

We will not always succeed in keeping our circle closed. Whatever the obstacles, we can *state the intention* to close our boundaries *even if* we do not know how!

When I recently presented the rope exercise in a workshop, several of my students had different reactions to setting boundaries. I then drew up a series of questions about their reactions to boundaries to clarify their positions. Here they are:

1. If the person is feeling better and less isolated when her/his boundaries are open, TEST if the person is making a mental, emotional or spiritual association between having boundaries and being isolated.
2. If the person did not feel a difference between both states, TEST if the person can clearly perceive her/his boundaries.
3. TEST if the person has in her/his genetic code a program to clearly perceive her/his boundaries.
4. TEST if the person thinks that she/he has the right to have boundaries.
5. TEST if the person thinks that she/he can be loved and be in a relationship if she/he defines boundaries.

CHAPTER 10
PROJECTIONS AND INTERFERENCE

HELPING VERSUS SUPPORTING

When our children, our friends or our spouse are not doing well or are bored, instead of trying to help them, we could simply recognize the situation and ask them what they would like to do to improve the situation. None of our help or suggestions would be required here, just our unconditional support. Trying to solve other people's problems is not supporting them. Our attempts at helping them only open up our own boundaries and lead us to invade their circle. By supporting others, we affirm our conviction that they are strong and powerful, and that we believe they essentially have everything they need to deal with the current problem.

Of course, it is much easier to support our clients than those close to us. Within a family, the circles are closely intertwined and family members, because of their genetic similarities, behave like compatible computers. In such a technological network, files can be transferred wirelessly between computers with the BlueTooth transmission technology. In general, we may even consider that we need to be separated from one another to maintain our individuality. In fact, some of us prefer to be alone rather than in a relationship, out of fear of such a transfer.

And yet, by maintaining our boundaries at zero point, we can have our cake and eat it too. We can *choose* to have intimate relationships and keep our individuality *even if* we do not know how. In order to properly understand this concept, we have to look at the law of entrainment and at projections.

LAW OF ENTRAINMENT

We all know couples in which one of the spouses is experiencing spiritual, artistic or another type of advancement while the other appears to be going nowhere. This was the case for one of my friends. When we looked at the situation without judging, we realized that my friend was polarized. Her rhythm of evolution was **right** and that of her companion, **off**. Now, if we apply the quantum model to this observation, we see that it is defying the standards of matter because, according to this model, everything coexists at different vibrational frequencies. There is no superior or inferior partner in the quantum model. My friend's evolution and the so-called 'slowness' of her spouse are both part of the quantum flux. Acknowledging this enabled us to open a gap in the standard model of intimate relationships.

We analyzed the situation based on the principle of how a pendulum works. In physics, an extraordinary law exists, the **law of entrainment**. It was discovered in the 17th century by a European inventor by the name of Christian Huygens. In accordance with this law, when magnetic forces or fields are aligned, information flows among them. If two elements are next to each other and vibrate at different frequencies, the one that vibrates more slowly will tune its frequency to that of the other through the effect of entrainment. For example, if we place several pendulum clocks in the same room, the slower pendulums, as though under the effect of an irresistible influence, will slowly begin to oscillate at the same rate as the faster pendulums.

Going back to my friend's situation in her relationship, we asked ourselves why she needed her spouse to adapt to her rhythm. Who said that evolving quickly was better? And finally, why should we adapt ourselves to the fastest cruising speed? From the perspective of zero point, this rule appears excessively polarized. Quantum physics has shown that certain rules cannot be proven in all situations. Might this be the case for the law of entrainment? Since everything coexists, how can we say that my friend's rhythm is more **divine** than that of her spouse? The quantum model, with its concept of coexistence, does not necessarily confirm the law of entrainment, whereby the slowest is influenced by the fastest. The quantum model offers multiple possibilities that fluctuate incessantly, without one being any more true than another. My friend and I understood that we needed to find a way to impede the law of entrainment. What bothered her was not the rhythm of evolution of her spouse, but the fact that she was entrained by his rhythm, which was slower than her own. In the same way, her spouse was irritated at having to follow her faster rhythm.

Following a brainstorming session, we found the following emotion: *desire to be controlled*. What a surprise to find that not only do we want to control the other person's rhythm, but that in the core of our being, we also want to be controlled. We are torn between contradictory movements. Even if we understand that their own frequencies are appropriate for them, we want our loved ones to vibrate at our frequency rate so as not be entrained by theirs; or, we want them to vibrate at a frequency higher than ours to allow ourselves to be entrained to vibrate at a faster frequency.

When we become responsible for our frequency rate, we understand that it is better to maintain our essence and our individual rhythm because the wavelength of others is as foreign to us as ours is to them. Unconsciously, we will do everything in our power to resist the law of entrainment: we might quarrel, cross swords, compete, criticize, attempt to persuade the other and so on, to be able to maintain our own frequency, control that of others or be drawn into their rhythm.

PROJECTIONS

It is easy to lose sight of ourselves and to project scenarios onto others rather than remain present to our own personal history. This is what happens when we judge others instead of understanding our own needs and wounds. The best way to interrupt this vicious circle is to redirect our observations toward ourselves, in a compassionate fashion.

I have noticed that many couples have vibrational affinities, but also similar wounds. Just like in parent-child relationships, because of transgenerational memories or traumas experienced as children, their wounds resonate between them and create feelings of discomfort when the spouses are together. Not only do we get married for better and for worse, but we are also mutually attracted by what is best and worst in us. It is possible that we detect our spouse's wound and that this wound awakens our own without our knowledge. In the presence of our spouse's wound, we resonate to the uncomfortable effects of this unconscious wound and become irritated, tired and defensive. In this case, one of the spouses, the first one to detect the wound, has to break this chain reaction. It is not such a bad thing to be the one who is responsible for interrupting the pattern and making this first move. The spouse who does so becomes the agent of change. When we realize that old wounds play an important role in the manifestation of our reality, it becomes easier to accept this responsibility.

In my practice, I have had the privilege of meeting many parents who have consciously decided to no longer blame their children and rather work on themselves than on the latter. They have given up criticism and, instead, are doing conscious work on themselves. In doing so, they have obtained spectacular results with their children.

The wounds of two members of the same family or two spouses can resonate electromagnetically from one brain to another. This resonance resembles a wave in the shape of an '8' that passes from one brain to another. I have even met spouses who can trigger attacks of hypoglycemia in each other in periods of stress.

When we criticize our spouse, our criticism always has to do with childhood experiences. In fact, we choose our partner precisely because that person has gone through similar experiences. Having a spouse brings us back to moments of intimacy, family, childhood or moments in our lives when we were young and powerless in a world of adults. Using kinesiology or another method, we can determine the age of these initial wounds that are awoken by our companion's behavior. With our adult perspective, we can return to the wound consciously, unravel this moment of powerlessness and give it a new intention. The wound will never disappear, but through intentions we will use it differently. These conflicts are stimulated by the context of the intimate relationship and not by the person we love. Once again, we can *choose* to love ourselves within this intimate relationship *while* accepting that our wound has been awoken by our spouse. We will stop judging ourselves and others and we will love ourselves *while* knowing that we and the other are not perfect.

Sometimes our criticism of others conveys the discomfort we feel when we sense that the other person does not vibrate at the frequency rate of her/his essence. The other person behaves like a mirror and this is unbearable: we want that person to be aligned with her/his essence so that there are no more boundaries, or limits of time or space. We are criticizing the reflection of our own limits because we then feel very far from our essence. Yet, its essential vibration is essential to us.

Intentions must always be stated independently of other individuals' names. If we state an intention that involves a specific person, we place ourselves in a position of self-defense, control and interference. Then, our intention is not at zero point and it no longer respects the free will of others. By stating an intention at zero point and by creating a new program, I define new boundaries. These boundaries are healthy because they emerge from my own intention **about myself** and constitute my own new choices. If, on the contrary, my intention concerns someone else, I place myself in a defensive position, which implies that I have cause to defend myself. However, this impression can be brought about uniquely by my unconscious programs. Healthy boundaries established by means of an intention at zero point spare us from having to defend ourselves because, by including our fears, they become extremely effective. This is exactly like in martial arts: the real masters almost never fight because their assurance and power erect boundaries that others feel and do not cross.

There is no need to be afraid of intimacy if we favor the new quantum intimacy, which advocates coexistence. We can allow the other to feel discomfort in our presence and allow ourselves to feel our own discomfort inside our closed circle.

We can feel at ease in our family, our couple, at work or in a group by supporting others instead of rushing to help them and by granting them the time they need to complete their process in their own way. Compassion requires that we let others live their lives and that we give ourselves permission to doubt and understand things at our own rhythm while trusting in ourselves *even if* we do not know how. By being self-supportive, we support others and make every experience something sacred. We accept that our loved ones awaken our old wounds and take it upon ourselves to become aware of our unconscious and codependent programs, *even if* we do not know how. We follow our own process and we do so according to our own rhythm. When we disassociate ourselves from the survival and adaptation process of those close to us, we progressively claim our personal frequency rate and we progress toward conscious ascension.

Interventions Without Interfering

I am convinced that we suffer from a form of collective dependency that consists of interfering in other people's affairs or wanting others to interfere in our own affairs and save us.

The easiest way to understand the concept of boundaries is by refraining from interfering with other people's vibrations. We will then have a better understanding of the nature of a healthy boundary. If we try to master boundaries, we risk falling into perfectionism and going to extremes, while building walls rather than boundaries. By learning to respect the vibrational rhythms of others, we will automatically understand how to manage our own boundaries. When we interfere in the affairs of other people, we take up their vibrational space and block their own access to this space. By withdrawing from their quantum path and allowing them to vibrate at their own frequency rate, we allow them to initiate movement in their lives. Then, instead of criticizing others because they do not move or change their oppressive behavior, we support them without trespassing on their vibrational space so that they can occupy it fully on their own.

It is normal to want to help someone who is suffering, but there has to be some sort of vibrational resonance. Our approach has to be in harmony with their frequency rate. It is possible that our rate may not suit them. We can declare our availability while leaving them free to take advantage of it or not. I am not talking here about an emergency or a trauma, but about more ordinary circumstances. Our desire to help is seldom at zero point and could stem rather from a need for recognition. By becoming conscious of the trap that interference represents, we will have our willingness to help coexist with our desire to be recognized without invading the space of other people.

One of my clients was wondering how she could nurture her children with healthy boundaries. A baby needs to be mothered and fussed over. But when we mother older people, we tend to overprotect them. The difference between the two is a question of open or closed boundaries. Compulsive mothering is an internal drive that, in the case of my client, was pushing her to interfere in the lives of others. Deep down, this compulsive desire is not about nurturing others but about having wide open boundaries. My client had unconsciously learned that healthy boundaries meant rejecting others. We tend to believe that if we close our boundaries, we will be cut off from those close to us and we will be isolated. We may believe that we will no longer have the right to be loved, will no longer be cherished, cuddled or supported or will no longer be able to treat those who are close to us generously. On the contrary, healthy boundaries assure authenticity in our nurturing of others.

If our essential nature pushes us to mother others, then we will be even better nurturers if our boundaries are closed. By choosing to bring ourselves back to ourselves and amplify the vibration of our essence, we will be able to take care of others without wearing ourselves out or feeling drained.

When people pray for us without our consent, they could very well be interfering in our lives. Later on, if we will feel better, they might claim that it was their prayers that worked instead of our own intentions. In our moments of weakness or during important transitions, we can look for support, ask others to pray for us and accept to receive from others. We are not alone in the world and can feel that we belong to our community when we know others can pray for us. However, in the support paradigm versus the helpful approach we are co-creators and partake in our well-being. We can take part in our own healing. The best prayer that I know is the following: *Infinite support comes in infinite ways.*

If someone around us refuses our prayers or the therapeutic help that we offer, we should observe our reactions to see whether we are interfering or not. If we accept that person's attitude with calm assurance, we will know that we are finally free from trespassing boundaries.

There are all sorts of seemingly good enough reasons for interfering in the affairs of others. For example, we might consider that our prayers are effective or that we have received a message to intervene by contacting the person's soul or mental body without that person's conscious permission. It is important to acknowledge that, apart from the physical body, we also have emotional, mental and spiritual bodies. When we insist on imposing our helpful interventions without a person's conscious awareness, we might not be able to fully acknowledge the specific requirements of the four bodies and the person's level of consciousness. We might misunderstand the person's individual vibrational requirement and impose our own frequency.

The Various Frequency Rates of Our Interventions

Imagine a painting with a series of light rays that each vibrate at a different frequency. One of the rays vibrates at the frequency rate of bikers, another at the frequency rate of drama students, the following at the frequency rate of mothers who have just given birth, etc. Likewise, there are all sorts of people on our planet. Sometimes, our parallel lives cross one another but, most often, they simply coexist.

Different therapeutic approaches also vibrate at specific frequencies. When we are trying too hard to help someone, we may be imposing the vibrational frequency rate of a ray that is not in harmony with the frequency rate of this person. A biker who is single will not respond to the same approach used for a mother who has just given birth. Nevertheless, according to the laws of coexistence, no ray is better than another–everything is a question of choice and resonance.

The era of saviors is out. Now is the time to work on maintaining a frequency rate. When I maintain the frequency rate of self-love and peaceful assertiveness, I emit a vibration with which other people who vibrate at a similar frequency rate can resonate. The time for saving others is over. The time for healing is over. Now is the time for specific frequencies. Our work consists in maintaining the frequency rate of our essence and self-love *even if* we do not know how.

In the near future, instead of speaking about others, about what they do in life and what they own, we will be able to explain how we have managed to keep the vibrational frequency rate of our essence intact in their company. Conversations will become so much more interesting. Instead of speaking of vibrational disharmony, we will speak of tonalities that are perfectly attuned, regardless of the situation. We will no longer speak about others but about ourselves. We will no longer speak about our wounds but about the road we have traveled to achieve the full manifestation of our individual essence. Our spirituality will become that of the essential Unity. Our story will become that of our vibrational harmony with others and with ourselves.

When individuals vibrate at the frequency rate of their essence, they cannot be manipulated. They can no longer play the role of victims or be invaded by others. It is impossible because they no longer vibrate at the frequency rate of others. They will feel invaded only when they find themselves on a path other than their own.

We have never managed the amazing feat of maintaining our frequency rate within a polarized physical world. Today, because of the new paradigm of zero point, we have the opportunity to live this experience and to be the driving force behind this revolution. We must not try to adjust our frequency rate to save as many people as possible. On the contrary, we have to preserve our integrity and maintain the frequency rate of our essence while allowing others to exist alongside us at their own personal rate. This coexistence is the profound expression of our mutual well-being. The goal of existence is to **feel**

good about ourselves and allow others to exist and to vibrate at a frequency rate that suits them *while* **feeling happy in their vibration!** As long as we seek to save others, we do not really feel good about ourselves; we are polarized and have strayed from zero point. Our circle is open; we are not connected to our frequency rate–we interfere in the lives of others to influence them to tune in on our frequency rate instead of being content to emit our own frequency.

This concept of interference is so subtle, so insidious and so paved with good (polarized!) intentions that I have developed an exercise that will help us to detect and integrate it.

This exercise constitutes a crucial step in the process of self-love and healthy boundaries. To ascend, all aspects of our being have to, among other things, vibrate at the frequency rate of our essence. How can we manage to do this if we are busy invading the space of others with our own frequency? If we stay in our own space, we will preserve our vibrational integrity.

When he spoke of ascension, Jesus said that what he had done, any man could do: he was encouraging us to identify with the vibrational wave of our essence to be able to ascend. Unfortunately, the meaning of his message escaped us and we tend to think that we have to vibrate at the frequency rate of Christ to be saved. If the frequencies of Krishna, Jesus or Buddha were the only ones that existed, how dull the world would be. Each of us has a remarkable individual frequency. By clarifying our vibrational space using healthy boundaries, we will manage to manifest this uniqueness. Imagine the pleasure of an encounter in which all involved emit their own frequency rates without influencing those of the others. Unconditional love is born from such authentic vibrational encounters. We will become drops in the ocean, defined by our essential nature, able to receive from the Whole and give back to the Whole. Healthy boundaries combined with intentions at zero point will open wide for us the doors to the ancient alchemical model of Unity where the One merges with the All and remains unique.

Take the time to ponder and reflect on the following questions. Intuitively answer by yes or no, or use other intuitive means such as kinesiology or the pendulum to get your answers. These questions were designed to induce reflection and will allow you to deepen your thoughts.

Exercise 2
MAKING INTERVENTIONS WITHOUT VIBRATIONAL INTERFERENCE

You can use this exercise on yourself or as a therapeutic tool.

After answering the following questions and establishing your individual profile, you can choose among various tools such as prayer, meditation or DNA reprogramming to put the situation back at zero point.

If you choose DNA reprogramming, test all of the following questions to determine whether you obtain a yes or a no. Then, test to determine how many protocols you need to put these questions at zero point. (See DNA Demystified *and* DNA and the Quantum Choice.*)*

1. TEST if the person can have her/his own individual essence vibrate while helping someone else:

A. without interfering in the other person's life;
B. without feeling responsible for the outcome of the other person's process;
C. without feeling guilty regarding the outcome of the other person's process;
D. at zero point without seeking to convince the other person;
E. without wanting to direct the other person to reach an outcome that is not her/his very own;
F. with the other person's conscious permission;
G. without taking up the other person's vibrational space;
H. while staying in her/his own vibrational space;
I. while maintaining her/his integrity at the following levels:
- **a.** emotional
- **b.** mental
- **c.** spiritual
- **d.** vibrational
- **e.** other

J. at zero point without competing with the other person;
K. other.

2. TEST if the person can have her/his own individual frequency vibrate while counseling someone else:

A. without interfering in the other person's life;
B. without feeling responsible for the outcome of the other person's process;
C. without feeling guilty regarding the outcome of the other person's process;
D. at zero point without seeking to convince the other person;
E. without wanting to direct the other person to reach an outcome that is not her/his very own;
F. with the other person's conscious permission;
G. without taking up the other person's vibrational space;
H. while staying in her/his own vibrational space;
I. while maintaining her/his integrity at the following levels:
- **a.** emotional
- **b.** mental
- **c.** spiritual
- **d.** vibrational
- **e.** other

J. at zero point without competing with the other person;
K. other.

3. TEST if the person can have her/his own individual frequency vibrate while intervening with someone else:

A. without interfering in the other person's life;
B. without feeling responsible for the outcome of the other person's process;
C. without feeling guilty regarding the outcome of the other person's process;
D. at zero point without seeking to convince the other person;
E. without wanting to direct the other person to reach an outcome that is not her/his very own;
F. with the other person's conscious permission;
G. without taking up the other person's vibrational space;
H. while staying in her/his own vibrational space;
I. while maintaining her/his integrity at the following levels:
- **a.** emotional
- **b.** mental

  - **c.** spiritual
  - **d.** vibrational
  - **e.** other
- **J.** at zero point without competing with the other person;
- **K.** other.

4. TEST if the person can have her/his own individual frequency vibrate while nuturing someone else:

- **A.** without interfering in the other person's life;
- **B.** without feeling responsible for the outcome of the other person's process;
- **C.** without feeling guilty regarding the outcome of the other person's process;
- **D.** at zero point without seeking to convince the other person;
- **E.** without wanting to direct the other person to reach an outcome that is not her/his very own;
- **F.** with the other person's conscious permission;
- **G.** without taking up the other person's vibrational space;
- **H.** while staying in her/his own vibrational space;
- **I.** while maintaining her/his integrity at the following levels:
  - **a.** emotional
  - **b.** mental
  - **c.** spiritual
  - **d.** vibrational
  - **e.** other
- **J.** at zero point without competing with the other person;
- **K.** other.

5. TEST if the person can have her/his own individual frequency vibrate while listening to someone else:

- **A.** without interfering in the other person's life;
- **B.** without feeling responsible for the outcome of the other person's process;
- **C.** without feeling guilty regarding the outcome of the other person's process;
- **D.** at zero point without seeking to convince the other person;

E. without wanting to direct the other person to reach an outcome that is not her/his very own;
F. with the other person's conscious permission;
G. without taking up the other person's vibrational space;
H. while staying in her/his own vibrational space;
I. while maintaining her/his integrity at the following levels:
  a. emotional
  b. mental
  c. spiritual
  d. vibrational
  e. other
J. at zero point without competing with the other person;
K. other.

6. TEST if the person can have her/his own individual frequency vibrate while reprogramming someone else's DNA:

A. without interfering in the other person's life;
B. without feeling responsible for the outcome of the other person's process;
C. without feeling guilty regarding the outcome of the other person's process;
D. at zero point without seeking to convince the other person;
E. without wanting to direct the other person to reach an outcome that is not her/his very own;
F. with the other person's conscious permission;
G. without taking up the other person's vibrational space;
H. while staying in her/his own vibrational space;
I. while maintaining her/his integrity at the following levels:
  a. emotional
  b. mental
  c. spiritual
  d. vibrational
  e. other
J. at zero point without competing with the other person;
K. other.

7. TEST if the person can have her/his own individual frequency vibrate while being in a relationship with someone else:

A. without interfering in the other person's life;

B. without feeling responsible for the outcome of the other person's process;

C. without feeling guilty regarding the outcome of the other person's process;

D. at zero point without seeking to convince the other person;

E. without wanting to direct the other person to reach an outcome that is not her/his very own;

F. with the other person's conscious permission;

G. without taking up the other person's vibrational space;

H. while staying in her/his own vibrational space;

I. while maintaining her/his integrity at the following levels:

- **a.** emotional
- **b.** mental
- **c.** spiritual
- **d.** vibrational
- **e.** other

J. at zero point without competing with the other person;

K. other.

8. TEST if the person can have her/his own individual frequency vibrate while working with others:

A. without interfering in their lives;

B. without feeling responsible for the outcome of their processes;

C. without feeling guilty regarding the outcome of their processes;

D. at zero point without seeking to convince them;

E. without trying to impose an outcome that is not their own very own;

F. with their conscious permission;

G. without taking up their vibrational space;

H. while staying in her/his own vibrational space;

I. while maintaining her/his integrity at the following levels:

- **a.** emotional

- **b.** mental
- **c.** spiritual
- **d.** vibrational
- **e.** other

J. at zero point without competing with them;
K. other.

9. TEST if the person can have her/his own individual frequency vibrate while taking care of someone else:

A. without interfering in the other person's life;
B. without feeling responsible for the outcome of the other person's process;
C. without feeling guilty regarding the outcome of the other person's process;
D. at zero point without seeking to convince the other person;
E. without wanting to direct the other person to reach an outcome that is not her/his very own;
F. with the other person's conscious permission;
G. without taking up the other person's vibrational space;
H. while staying in her/his own vibrational space;
I. while maintaining her/his integrity at the following levels:

- **a.** emotional
- **b.** mental
- **c.** spiritual
- **d.** vibrational
- **e.** other

J. at zero point without competing with the other person;
K. other.

10. TEST if the person can have her/his own individual essence vibrate while [another feeling or occupation to be identified in this instance (loving, conducting an orchestra, teaching, etc.)]. This emotional state or occupation requires that the person be in a relationship with someone else. If you do not have another example, proceed to the next point.

A. without interfering in the other person's life;
B. without feeling responsible for the outcome of the other person's process;
C. without feeling guilty regarding the outcome of the other person's process;
D. at zero point without seeking to convince the other person;
E. without wanting to direct the other person to reach an outcome that is not her/his very own;
F. with the other person's conscious permission;
G. without taking up the other person's vibrational space;
H. while staying in her/his own vibrational space;
I. while maintaining her/his integrity at the following levels:
  a. emotional
  b. mental
  c. spiritual
  d. vibrational
  e. other
J. at zero point without competing with the other person;
K. other.

11. TEST if the person has the genetic program required to:
A. understand others;
B. know that the experience of others can be different from her/his own;
C. accept that the experience of others can be different from her/his own;
D. be at zero point in respect to this difference;
E. let the other person live her/his experiences at her/his own rhythm without judging that person;
F. accept the speed at which the other person progresses;
G. accept the frequency rate of the other person;
H. accept the difference between her/his own vibrational frequency rate and that of the other;
I. accept the other's essence;
J. other.

12. TEST if, during an intervention, the person's conclusions are at zero point, if they adhere to the other's reality or if they are polarized and more appropriate to the person's own experience than to that of the other.

13. TEST if you need to redo the previous steps while replacing "someone else" with the name of a specific person among those around you.

Imagine, for example, that you work in an office for a boss who has a powerful influence over his team. As soon as you enter the office, you automatically give him permission to manage your vibrational space. The goal is to succeed in being a member of this work team without having to change your frequency rate so as to adjust to that of the boss. You might want to redo this exercise using your boss' name. Here is what one of my students wrote to me following this exercise:

"On the Sunday night following the workshop, I had quite a peculiar dream in which someone was expressing violent anger. I did not understand the meaning of this dream right away and I found it very disagreeable after the wonderful weekend I had just spent.

The next morning, I learned that my sister-in-law had died at about 11:30 p.m. the night before. I understood right away that she was the one who had appeared to me in a dream. I did not know her very well because she was living far away. I told my dream to my spouse who said to me, "After all she has been through, I would not be surprised that she should feel so angry."

Not long after, I had the impulse to play Ave Te Deum for my sister-in-law. I let my tears flow. They were not tears of sadness. Although I had known her very little, I was overflowing with love for her. I would even say that the love I felt was proportional to the anger that I had felt from her in my dream. This love seemed to be exactly what she needed at that moment. Unlike the other times I had felt strong emotional feelings of either joy or love within myself, my body did not begin to tremble. Neither my body nor my thoughts were troubled for the rest of the day. I experienced this moment peacefully while knowing that life went on and that I was not merging into the other. We were two distinct entities that had shared an unforgettable moment…

I don't know if it was an accident (actually, I don't believe in accidents!) or if the protocols have already begun to have an effect, but this experience was much easier to go through than usual.

Thank you Kishori and thank you, my Essence!"

Dominique T.

CHAPTER 11
PERSONAL NEEDS

To experience self-love, we have to return to our essence, and to recognize our essence, we must know ourselves. One way to get to know ourselves better is to know our personal needs. Each person has individual needs. Needs are as unique to a person as that person's vibrational frequency. When I say **need**, I am not talking about desire. A desire is a longing to obtain or possess something whereas a need is a natural necessity.

If needs vary from one individual to the next, our response to our personal needs is often similar. Most of us tend to question our emotional needs and reactions. In addition, advertising and fashion dictates tend to cut us off from our physical needs by imposing false needs and providing false answers instead of showing us how to listen to our body and follow our innate intelligence.

Humans not only have physical and emotional needs, but intellectual and spiritual needs as well. These are not as palpable, but we simply have to remind ourselves to what extent a life devoid of inspiration is boring to conclude that it is important to satisfy these personal needs as well. Why be ashamed of our emotional needs, such as the need for recognition and fame, for example? We have a light frequency that is completely individual and unique. This unique frequency should be acknowledged. Most of our personal needs are unconsciously born of our desire to identify with our essence. They are not there to humiliate or degrade us, but to give us an identity of our own.

To develop properly, children have to be treated appropriately and live in a context adapted to their natural dispositions. This also applies to adults who continue to experience the four types of personal needs listed above. The difference lies in the way in which we satisfy these needs. Many of my clients complain that their spouse, their friends or their colleagues do not take care of their personal needs. However, I have also noticed that when people make sure their own needs are fulfilled in a healthy and effortless fashion, the people around them fully support their approach. Satisfied individuals seem to attract support and sympathy. People who do not take care of themselves, who do not know themselves, who do not manage to satisfy their personal needs and who rely on those close to them to do the job, will experience conflicts and problems. Expecting others to take care of us in our stead can only bring about resistance and snags. Others are faced with the same personal challenges and have to clarify their intentions and choose to fulfill their own needs regardless of their personal situations. Just like us, they are faced with lack of self-love, with default programming and with a feeling of powerlessness. How can they take charge of our lives and be meeting our needs? They have enough taking care of their own. Basically, all they can do is offer their support in our effort to meet our own needs.

Indeed, the support of others confirms that we have the capacity to take care of ourselves. The only way to put this theory into practice is to *state the intention* to take care of our personal needs *even if* we do not know how. Otherwise, it is preferable that we avoid throwing ourselves into new projects because we will always be waiting for someone else to take care of us. Next time, before making a new decision, we should stop for a moment to be sure that the initial intent takes our personal needs into account. We should not be ashamed of our personal needs; rather, we should understand that they constitute an asset and contribute to our success.

It is essential that we pay constant attention to our personal needs. Some parents only help their children with their homework every two weeks or occasionally. These children do not receive constant support and it is certain that they will have poorer marks. Providing steady attention on a daily basis may, despite our changing needs and our emotional wounds, instill a state of inner peace and a sense of stability. Then, when appropriate, we can seek out professional help to support a difficult process without becoming codependent. We will no longer look for miraculous healing and will be freed of our perfectionism. Gradually, we will develop a reflex that will flout the guilt associated with being an imperfect person with specific needs.

Taking care of our personal needs should not be a stressful endeavor. It would be easy to meet our needs if we accepted who we are and acted with compassion toward ourselves. It is more efficient to choose to attune ourselves to the tonality of our essence than to endlessly elaborate new stratagems for taking our personal needs into account. Furthermore, we can assist this process by including the negative polarity of our disappointments or the obstacles that stand in our path. On days when we have difficulty taking care of our personal needs, it is by accepting our inability to be present to ourselves that we can keep this process flowing. If our focus is directed toward self-acceptance, taking care of ourselves will become a natural state rather than an extraordinary event.

The path to consciousness on Earth is full of pitfalls. By disregarding our personal needs, we may easily give up on the process. We are human and, at times, our experience on Earth is no picnic. To succeed, we have to learn to respect and include what is natural and necessary. Physical and emotional hardships can lead us to give up; we may become discouraged and let go of our spiritual desire to ascend. By developing an attitude at zero point in which our spiritual side, devoid of human needs, coexists with our very carnal side, we will succeed in including everything we are in a single conscious process. This is why I encourage you to dive into your deepest depths to discover those instinctual personal requirements that enhance your individuality to a certain extent.

To help us better identify the way we meet our personal needs, I have drawn up a list of questions to be answered by yes or no. Above all, do not judge your answers. The goal of this exercise is simply to establish a portrait of your reactions. Answers to this questionnaire vary from one person to the next, but each individual can reach some personal understanding from the answers. The exercise will help you recognize your personal needs in respect to a particular project or situation. You will then be able to formulate a powerful intention.

Exercise 3
IDENTIFYING OUR PERSONAL NEEDS

You can use this exercise on yourself or as a therapeutic tool.

After answering the following questions and establishing your individual profile, you can choose from various tools such as prayer, meditation or DNA reprogramming to put the situation back at zero point.

If you choose DNA reprogramming, test all of the following questions to determine whether you obtain a yes or a no. Then, test to determine how many protocols you need to put these questions at zero point. (See DNA Demystified *and* DNA and the Quantum Choice.*)*

You can apply this exercise to a particular situation. If so, identify the situation.

TEST if, when taking care of her/his own specific needs, the person:

1. thinks she/he is going to die;
2. finds it demanding;
3. knows that she/he exists;
4. knows that she/he is important;
5. feels blessed;
6. can experience expansion while respecting her/his personal needs;
7. knows how to meet her/his personal needs simply and effortlessly;
8. knows that if everyone meets her/his own personal needs, the result will be harmonious;
9. can take care of her/his personal needs continuously and permanently;
10. can set goals that take into account her/his personal needs;
11. is faced with a physical, emotional, mental or spiritual block;
12. can remain open to others;
For example: When asked to do something the person does not want to do, the answer could be, "Your idea is very appealing, but this may not be a good time for me."

13. can take pleasure in taking care of her/his personal needs;
14. has difficulty accepting that others take care of their own needs whereas she/he does not do so;
15. takes care of the needs of others so as not to have to take care of her/his own;
16. can feel self-love when taking care of her/his personal needs while:
  a. holding a grudge;
  b. sulking;
  c. feeling sad;
  d. being angry;
  e. feeling isolated;
  f. re-living violent, abusive or other events (from what age?);
  g. leaving the here-and-now;
  h. being afraid of lacking something;
  i. justifying herself/himself;
  j. being sick;
  k. suffering from a hormonal imbalance;
  l. suffering from a structural imbalance;
  m. suffering from another imbalance;
  n. attributing the same needs to others (transfer);
  o. being stressed;
  p. feeling self-hatred;
  q. feeling ashamed;
  r. being hurt.

TEST if the person can take care of herself/himself when someone else is needy:
1. without feeling guilty;
2. without feeling that she/he is abandoning the other person;
3. while knowing that she/he has done all that could be done;
4. while supporting the other person rather than helping;
5. while knowing that she/he has given all that could be given;
6. without criticizing the other person's needs;
7. while knowing that she/he can refuse to take care of the other person's needs;

8. while knowing that she/he can take a moment before agreeing to the requests of others without feeling guilty;
9. while knowing that her/his own vibrational approach may not be aligned with the person's frequency.

Intentions and Personal Needs

My students and I have noticed that the power of intentions at zero point is exponential. Intentions at zero point have formidable effectiveness. The most complicated part is not to formulate an intention at zero point, but to determine the right intention. I now understand that the secret to a good intention lies in the depths of our understanding of our personal needs. From now on, we will voice intentions based on our needs.

A client who wanted to take up a career as an artist threw herself into this path with all her passion and creativity. After a certain time she had to put the brakes on her project. She was tired, could not manage to make ends meet, had no more time for herself, and so on. Even the best project is doomed to slow down or fail if it does not take our personal needs into account.

According to certain philosophies, our personal needs hinder our evolution. Yet, we are human and it is as such that we will live the experience of conscious ascension. Why fear our humanity? We have come to Earth to experience duality, put it at zero point and learn to command what we want with it. From the moment that we understand how to make the most of the destructive patterns of duality, we are no longer enslaved by these patterns. Human beings who are well rooted in their essence use their personal needs to confidently express themselves within a world made of duality. The current planetary frequency rate can lead us to adopt the frequency rate of love in duality, not to deny it. This is the new spirituality. It has to do with the manifestation of a new frequency rate. The new paradigm consists of integrating binary thinking and entering zero point, of accessing our unconscious blocks by giving them their rightful place and giving up our judgments.

Our goal is to develop a supportive and responsible attitude with regard to ourselves. As we have seen above, instead of helping supposedly weak and incompetent people, it is better to communicate our conviction that they will find the perfect solution for themselves, that we feel that they are competent in meeting their needs. We have to adopt the same attitude for ourselves. To break free from our emotional, physical, mental and spiritual codependencies, we have to access the power of intention and start proclaiming that today *we choose* to live in harmony with our personal needs *even if* we do not know how.

In the same way that hunger disappears after a good meal, our personal needs become invisible when they have been met. Thus, we must adopt a new attitude regarding our personal needs. A plant needs water and we would not buy any if we did not have the intention of watering them. At zero point, it is possible to have our tendency to censure our personal needs coexist with our self-commitment. Rather than having to meet our ever-changing needs, we could start developing a new consistent approach. Thus, instead of stating the intention to have time for ourselves or to be able to meet our personal needs, we would *choose to* live in a state of continual acceptance of ourselves and of our personal needs *even if* we do not know how.

By developing an ongoing relationship with our personal needs, we will automatically develop a new behavioral attitude. Suppose that one day I need to rest quietly: I will maintain the frequency rate of my essence in doing nothing. If the next day my agenda is full, I will maintain the frequency rate of my essence in action. Our relationship to our personal needs has to become as natural as our relationship with our individual essence. It is a bit like marrying oneself: we promise to love ourselves for better and for worse. We vow to cherish and honor ourselves for richer or poorer, in joy and in sadness, in health and in sickness. We solemnly promise to be our own support.

Take for example a student named Andrew. Every Monday morning, Andrew finds it very difficult to get back into student mode. He could *state the intention* that the return to his college become effortless *even if* he does not know how, but we could find a more powerful intention. This is why we made a list of what Andrew's needs are when he resumes his life as a student. Here it is:

to see my friends
to remain caught-up in my schoolwork
to be calm
to have energy

to have new experiences
to take initiatives
to learn
to use my creativity
to keep an open mind
to be inspired

Next, Andrew had to choose the need that seemed to be a priority for him. This being **inspiration**, he then stated the following intention, "*I choose* to be inspired when I am in college *even though* I feel bored." This intention clearly expresses one of Andrew's personal needs.

I redid this exercise with another client, Melanie, whose professional responsibilities had become overwhelming and, together, we drew up the following list of needs:

pleasure
freedom
health
balance

Melanie's most prominent needs were **freedom and health**. Melanie therefore stated an intention based on these personal needs and to which she gave priority, "*I command* or *I choose* health and freedom *while* being afraid of being overworked." She was then able to assess all the requests made of her at work and use this intention in accordance with her personal needs. Melanie noticed that she soon was able to achieve a greater level of productivity and pleasure.

From now on, we will reframe our choices by stating an intention based on our personal needs. Take, for example, the following intention, "*I choose* to be free and healthy *while* responding to many demands." When we voice an intention, we must let the form/effect go and not try to determine how it will manifest itself. If our intention encompasses our needs, we will find it easier to let ourselves be carried along without worrying. Of course, we all have inner concerns that our needs will not be met but ultimately, what worries us deep down is not to be self-supporting.

The next exercise consists in formulating an intention based on personal needs. For example, suppose that we are looking for an intention that will help us accomplish a project. It is important to identify our needs regarding this project, **even before coming up with our core intention**. To do this exercise correctly, you first have to clearly identify the situaation. Step back a bit, as though you were watching a film, and examine the situation carefully. Once you have figured out the context, establish the list of your personal needs.

People who live according to others and have been repressing their needs for years have difficulty determining their personal needs. Others feel a sense of urgency and would like to find the right intention once and for all. This exercise is not a test. It is simply a question of drawing up a list of our personal needs. We can come up with simple and easy answers. It is hard enough to stop and listen to our needs; we do not need to come up with complicated answers. This is not a school assignment; rather it is an intimate encounter with ourselves.

I invite you to appreciate the uniqueness of your personal needs. Welcome them with compassion, do not judge them and do not feel uncomfortable by the nature of your specific desires. We should not be ashamed of needing to feel safe or of looking for approval. Many doctrines criticize the need for approval. At zero point, *I can choose* to need approval *while* deeming this to be childish. The idea is not to improve ourselves to the point where we no longer express individual needs. We should not fear our simple needs such as the need for affection and comfort in our love relationship. We do not have to change ourselves; rather, we have to change our intentions.

The goal of this process is to succeed in ascending through the frequency rate of self-love. Ascension is the ultimate intimate experience, the complete acceptance of everything we are. To succeed, we will have to use, among others, the magnetic charges of our personal needs–whatever they might be–to reinforce our intentions. There is more power in identifying our personal needs and integrating them in our intentions than there is in trying to rid ourselves of them because of shame. We could not exist without needs. A plant needs water to live and we do not judge it for all that. Personal needs are part of our creative process. We will learn how to use them to manifest our essential identity. Identifying with our own frequency rate will then lead us towards ascension. It is worth taking such a risk! If someone tells you, “It’s scary how much you need approval!” answer by saying, "Thank you for bringing it to my attention; I will actually use this need to manifest what I need."

Before doing this exercise, take a moment to read the story of a client–let us call her Rosemary–who was getting ready to move. Rosemary is an organized and efficient woman. As the move drew nearer, she came down with a terrible flu and wondered what had caused it. Together we drew up the list of her personal needs:

to delegate
to remain flexible
to sleep well
to have moments of escape

We determined that **to delegate** was her priority need.

Rosemary understood that if she chose to delegate with confidence, even while being afraid to let go, she would be more flexible, would have more moments of escape and would sleep better!!!!

Exercise 4
IDENTIFYING OUR PRIORITY NEED

You can use this exercise on yourself or as a therapeutic tool.

1. DEFINE the project or situation to include in this session.
2. DRAW UP the list of your personal needs (you can consult the list that follows if you are short of ideas).
3. IDENTIFY the need that has **top priority**.
4. FORMULATE your intention by including your need *while*...
5. Once you have formulated your intention you can consolidate it using various means, such as prayer, meditation, or DNA reprogramming. If you choose DNA reprogramming, verify how many protocols you need to install to put your intention at zero point. (See *DNA Demystified* and *DNA and the Quantum Choice*.)

List of Personal Needs

1. To eat better
2. To drink more water
3. To learn – experience
4. To work
5. Career
6. Support
7. To belong to an association, group, etc.
8. Family
9. To have children
10. Love relationship
11. To play
12. To move - house
13. Regeneration - convalescence
14. Sexual intercourse
15. Treatment: acupuncture, medical, chiropractic, massage, surgery, psychology, skin care, etc.
16. To take initiatives
17. Inspiration
18. Belonging
19. Power
20. Feeling energized
21. To finish something - mourning, etc.
22. To delegate

1. Fulfillment
2. Competence
3. Approval
4. Ease
5. To assert oneself
6. Calm
7. Pleasure
8. Rest - to sleep
9. Vacation – to go on a trip
10. Health – well-being
11. Abundance
12. Balance
13. Safety
14. To be happy
15. To write
16. To laugh
17. Leisure – activities
18. To be creative
19. Change
20. Gratification
21. Success
22. The sea, mountains, water, the sun, oxygen
23. Other

PART III

ESSENCE

CHAPTER 12
THE FREQUENCY RATE OF OUR ESSENCE

In the previous chapter, we saw that when we are caught in a stagnant situation, it is because the situation is not vibrating at a frequency rate that is in harmony with our essence. Now that we know that everything is vibrational, we need only find out the vibrational rate of our individual frequency and how to tune in to it. To align ourselves along a quantum path that is in resonance with our essence, we must first have our essence vibrate within and then we must identify with it intimately. However, we are conditioned by other programs rather than by our own nature and we resonate more easily with the realities from our childhood than with the liberating frequency rate of our essence. *The Merriam Webster* Online dictionary defines **essence** as, "the individual, real, or ultimate nature of a thing especially **as opposed to its existence**." We tend to define ourselves by our unconscious programs rather than by our essence. These unconscious commands from our formative years come from our loved ones, our transgenerational bonds, our education and our emotional wounds. We tend to align our projects with these frequencies.

When I tried to understand the nature of my essence, I went back to memories of moments of happiness. Most of the memories that came to mind had nature and the forest as a backdrop. I noticed that in these snapshots of joy, I seemed to vibrate at a very special frequency rate. I understood that this was the frequency rate of my essence. I remembered moments from my childhood when, in spite of conflict, I came into contact with innocence. I have contacted this innocence time and again in my life. I have noticed that in general, I feel better and seem to vibrate more strongly when I am in the mountains or on water. I have a passion for mountains and water, and these two elements help me amplify the frequency rate of my essence. To have some understanding of the frequency rate of your essence, you have to clearly identify an environment in which you have the impression that you vibrate more strongly. Thus, a musician may have a kinesthetic experience of his frequency rate while playing music.

Therefore, to identify our essence, we have to first remember a kinesthetic moment marked by innocence. There are no set rules for determining these moments. They are different for each of us because we each have an essential signature that is our very own.

As I write these lines, the government is asserting increasing control and we are victims of the standardization of cultures and ways of life. Nevertheless, our essence remains unchanged in the face of external influences. It constitutes our intrinsic nature, independent of our environment. When I became aware of the incredible power that our essence represents, I understood that it was unique to each one of us. We all have a unique vibration that can be neither cloned nor copied. In a world marked by conformity, we can still enjoy free will because of our essence!

Of course, to recognize our essence, we need to know who we are. I am not talking about psychological decoding, since our essence is independent of our environment. In other words, if I am an apple, I am neither a banana nor a kiwi. I must accept that I am an apple. I will not allow my nature as an apple to be called into question because of a bruise. The bruise is one thing, but my essence is another and it is independent of this bruise. There is a very distinct boundary between the two. The suffering that we have experienced affects us but has no influence on our essence. The apple remains an apple whether we pick it or it falls from the tree.

We can judge our life in terms of failures and successes, or appreciate it without judgment in terms of our essence. Our understanding of our life is often shaded by a lack of self-esteem. To be proud of our essence, we think we should be successful. Yet our essence is independent of our accomplishments. We can choose to live our life on the defensive and afraid of external influences or in peace, trusting our essence. Essence is the quintessence of zero point. We need to know that even if our essence is independent from the place where we grew up or the place in which we now live, it incorporates everything we are. There is no part of ourselves that is not included in our essence. There is no hierarchy at this level. It is not a question of our essence having a very high energy level from which the other parts of ourselves are excluded. All aspects of our being are part of our essence. This is the new paradigm, the new definition of self.

When we are doing well, does our breathing stop? Of course not, and our heart continues to beat even when we are not doing well. These functions are constant, just like our essence. The sectarian mind and the duality of good and evil have nothing to do with essence. In quantum physics, matter is constant and behaves differently depending on the observer. We finally have a definition of who we are that goes beyond our wounds and transgenerational influences.

Owing to the quantum physics and quantum choice model, we can now go back and access information related to other possibilities that are in harmony with our essence. Indeed, according to the quantum model, the information from the choices we did not make in the past continues to exist in parallel worlds and we can have access to this information.

Take Thomas, for example. He wants to change careers. If he begins this new project without first accepting the fact that he dislikes his current job and that he thinks he made an error when he chose teaching, then his chances of creating a more dynamic situation in his life are reduced. Imagine that he has entered in his DNA the information relative to a career choice that is in greater harmony with his essence, *while* keeping his old choice. He will start his new career with a new outlook: the satisfaction he feels from having a profession that is more in tune with his essence will coexist with his feelings toward teaching. Consequently, he will start this new phase in his life differently: each of his steps, each of his commands will reach its goal with the speed and precision of an arrow. His projects will become reality instead of being hampered by the weight of recurrent models.

It will help Thomas to be conscious of his essence. This will provide him with the determination and energy needed to pursue his new path. He will base his new decision not on a childhood program but on his essence. When we make new choices based on our essence, we are finally making conscious choices that reflect our true nature.

Maintaining the Frequency Rate of Our Essence

We do not have to be mystics to tune in on the frequency rate of our essence. We can use the power generated by our emotions to maintain our intention. Then, we can reclaim the magnetic charge of these emotions and bring this charge into an intention that **aims to amplify the vibrational frequency rate of our essence in our interstitial void**. We can put all our energy into this wonderful goal and reinforce our frequency rate because our essence remains available as long as we exist.

The frequency rate of our essence is unique and vibrates at a specific rate.

We must be unconditionally devoted to the frequency rate of our essence. We cannot restrict it. Being inflexible can have a significant effect on our essence because it will confine it within restrictive parameters.

Because we long to be **saved**, we sometimes allow ourselves to be drawn along by the frequency rate of others. We unconsciously look for someone who vibrates at a faster frequency rate so we can tune our own frequency rate on to theirs. But another person's frequency rate is a very poor substitute for the vibration of our own essence. The same is true for the opposite process. When we try to convince someone to tune in on our frequency, we want to save that person. And yet, our frequency is never appropriate for the frequency rate of another person.

Making observations and discoveries about the unique and specific frequency rate of the essence of each person has given me true liberation. I was finally able to know who I was and to see myself from a perspective that went far beyond my transgenerational and emotional legacy. By visualizing happy moments when my frequency rate was high, I was able to recognize my unique **fragrance**. I then voiced the intention of having the frequency rate of my essence vibrate in my interstitial void both in my moments of solitude and in the presence of others. I was then able to amplify my vibrational frequency rate and diffuse it. Over time, I have developed greater self-confidence. Every day, I spend more and more time in this vibration and I am now becoming completely familiar with it. As I experience my unique vibration, I develop a greater sense of self-worth and harmony with the rest of Creation.

There is nothing surprising about this, since each entity in the cosmos also vibrates at its very own frequency rate while being part of the universal frequency. I am slowly letting go of the desire to control those close to me because I now know that others have their own particular essence that cannot vibrate at the same frequency rate as mine. I am happy to see them acquire their own identity and tune in on their own frequency. This experience, far from causing us to grow apart, has brought us closer together and increased the respect we have for one another.

The more I assimilate the quantum concepts, those of the Central Soul and of the essence, the more I perceive reality in terms of frequencies. My criteria for choosing food are no longer the same. I choose food not because it is organic, healthy or nutritional, but because it vibrates at a frequency rate that corresponds to that of my essence.

We waste so much energy trying to develop winning strategies. If we used this energy to maintain the frequency rate of our essence within ourselves, we would increasingly find ourselves in the right place at the right time and doing what is right for us. Our actions would become extremely effective because they would be governed by a powerful and limitless frequency, unique to each one of us. Harmony at zero point would take root within us. Of course, we would still experience difficulties and moments of depression. Our world would not become perfect overnight, but all the events in our lives would be in harmonious resonance with ourselves and we would not be anticipating events controlled by an unconscious idea of what life should be. We could finally manifest a life plan that is in resonance with our frequency and relax while carrying it out.

Through the power of intentions at zero point, we could use all the negative emotions that assail us to command a constant amplification of our essential frequency. Try it the next time you feel angry or demoralized. Breathe, feel your emotion, then command at zero point that its power amplify the frequency rate of your essence. Afterward, wait and see what happens around you in the hour that follows... It's worth trying–you'll be surprised! Gradually, as you keep experiencing your essence, you will find yourself in a state of consciousness similar to that of spiritual masters.

When masters enter an elevator packed with surly and disillusioned workers, instead of lingering on the state of mind of those around them, they maintain their individual frequency. This frequency does not need to be protected from the crowd. When these masters leave the elevator, their frequency rate has not changed one iota. It is precisely for this reason that we refer to these people as masters. If we show concern about the fact that they have been in contact with these people, they will answer that wherever they go, they are in paradise because they maintain the frequency rate of their essence, which is in direct contact with the Source.

With the string theory, physicists are studying a mathematical model that would appear to contain all the laws of physics. This mathematical structure redefines matter in terms of frequency rates. Now, I look at **everything** in terms of frequencies. I am beginning to experience within myself the discoveries of the quantum physics specialists regarding strings. According to the superstring theory, the form and content of the entire universe are determined by the vibrational frequencies of each atom. This concept corroborates the notion that, when all is said and done, there is no difference between matter and energy. I now understand that the realities of our lives are made up of thought-forms that vibrate at specific frequencies, and that this is true of all thoughts, even fear. Regardless of the situation in which I find myself, I choose to allow my individual frequency rate to vibrate. Increasingly, I feel that I am an essence vibrating in a human body even while this body is doing the dishes, for example. Doors have now opened not only about my life here, but also about parallel worlds.

The theory of multiple worlds stipulates that there are universes parallel to our own that vibrate at different frequencies. Indeed, research indicates that matter does not evolve solely in terms of width, height, depth and time. There are dimensions so minute that we cannot detect them, and so vast that we cannot access them. The April 2005 issue of *Science et Vie* describes this theory in the following way: "Minute identical strands, whose vibrations produce the diversity of our universe (...) The reality of our world, at its most basic level, appears to be a gigantic symphony resulting from minute identical strings... vibrating like the strings on a violin." [7]

Even if we do not always manage to maintain this frequency, we must not become discouraged. It is like learning to ride a bicycle. We will get to a point where, through regular practice, we will be able to be one with our essence. This process is exponential and circular. We will experience perfect moments followed by periods of doubt, and then we will come back to our essence with certainty. Gradually a network will be formed, but we have to expect accelerated movement. We will celebrate our successes and suffer great disappointment when we will have the impression that we have lost contact with our essence. This is all part of a non-linear process. Even the information that comes from choosing the wrong quantum path is important to help us maintain our frequency in the third dimension, which only obeys the laws of duality. The sadness associated with having vibrated at the wrong frequency is very powerful in the context of zero point. It will become the driving force of our decision and will give us the strength to continue.

7 Ibid., p. 58.

IDENTIFYING ESSENCE

To participate in the creation process by using the power of intention, we need to understand how important it is to know our own personal nuance. It is in amplifying our frequency rate through the power of intention at zero point that we will eventually learn to identify with our essence. To dare to be the creators of our lives, we need to first begin by accepting that we are unique and interesting and that we have a very distinctive flavor. This flavor is original and has a unique quality that cannot be cloned. It is this energy that we call our essence.

Our essence is our deep and individual nature, independent of our environment. It is our essential signature regardless of our domestic context and earthly embodiment. Even a child conceived in violence has more value than a cloned baby, precisely because of this child's essence. Despite the passion and tragedy surrounding its conception, this child has a uniqueness that the baby cloned in a test-tube, whose genetic legacy is supposedly perfect, does not have.

If we identify ourselves with our essence, we will finally know who we are, beyond what life has made of us. We will see ourselves as we are without the acquired models and defense mechanisms we have developed over the course of our existence. There have been Indigo Children born for several decades now. They are so different that educating them is a major challenge for their parents and teachers. In fact, what characterizes them is the magnitude of the frequency rate of their essence. They arrive on Earth with an essence that is more resonant, and they refuse to lose it despite medication and drugs. In addition, they possess a sixth sense that enables them to quickly recognize whether those around them vibrate in harmony or disharmony with their essence. In fact, the problem with these children is that their elders often have an essence that resonates less than theirs. If we could look at them without judging, we would understand that they are constantly encouraging us to be ourselves. They teach us to connect ourselves to our essence. Unlike us, they will never remember having had an essence with a low amplitude. They will not know that we needed blind faith to continue believing in ourselves.

Identifying with our essence gives us back our pride and self-love. The prouder we are of our essence, the less difficulty we will have with Indigo Children. By feeling passionate about our own frequency rate, it will be easier to coexist with individuals who differ from us. We will be able to vibrate, side by side, at our individual vibrational rates. By coming back to our essence, we will be able to strike up new types of relationships.

Our family had Indigo Children. Nowadays, Crystal Children are being born, and they are even more conscious, even closer to the vibrational rate of their essence. Through their presence, they force us to increase the amplitude of our frequency rate if we want to be able to understand them. By the time they are seven or eight, they have already understood the power of intention, can clearly express what they feel and know very well that we have to command what we want and not what we fear! These children and these young people are not better than we are: the difference has to do with personal vibrational amplification alone. All humans have an essence. We all come from the same Source. Over the years to come, if we choose to manifest our essence, we will no longer notice these differences.

Exercise 5
IDENTIFYING WITH OUR ESSENCE

You can use this exercise on yourself or as a therapeutic tool.

After answering the following questions and establishing your individual profile, you can choose from various tools such as prayer, meditation or DNA reprogramming to put the situation back at zero point.

If you choose DNA reprogramming, test all of the following questions to determine whether you obtain a yes or a no. Then, test to determine how many protocols you need to put these questions at zero point. (See DNA Demystified *and* DNA and the Quantum Choice.*)*

You can apply this exercise to a particular situation. If so, identify the situation.

TEST if, when identifying with the frequency rate of her/his essence, the person:

1. knows that life may undergo rapid expansion;
2. can command to be safe even if her/his life may change;
3. can command that the increased movement create harmony rather than chaos;
4. can feel greater vitality;
5. can feel the frequency rate of her/his essence inside;
6. can enjoy better health and feel physically revitalized;
7. can trust her/his essence;
8. can identify with her/his essence without falling ill as a result of the frequency changes due to the amplification of its vibrational rate;
9. can manifest reality on a quantum path that vibrates at the frequency rate of her/his essence;
10. can feel happiness, or any other emotion, in the quantum reality that vibrates at the frequency rate of her/his essence;
11. can recognize that her/his essence is a personal point of reference directly connected to the Source through the Central Soul;

12. can recognize that her/his essence is the anchor for the energy of matter and that it belongs to her/him through Creation;
13. can recognize that all of her/his incarnations carry the same essential signature;
14. can vibrate at the frequency rate of her/his essence without need for justification;
15. can meet her/his needs;
16. can establish a stable relationship with her/his essence;
17. can embody her/his essence physically;
18. can embody her/his essence on a daily basis;
19. can feel greater self-love;
20. can feel gratitude toward her/his essence;
21. can respect her/his essence;
22. can make choices in harmony with her/his essence;
23. can trust her/his essence;
24. can make wise choices that facilitate her/his evolution;
25. can stop worrying about her/his life process;
26. can follow a life-current that is fluid, consistent and compassionate, and in resonance with her/his own essence;
27. can participate in the movement of the universe while preserving her/his individuality;
28. can recognize or perceive her/his essence because of:
- **a.** its original nuance
- **b.** its unique tonality
- **c.** its effect
- **d.** its flavor
- **e.** its personal rhythm
- **f.** its vibrational frequency rate

29. can vibrate at the frequency rate of her/his essence:
- **a.** at work
- **b.** socially
- **c.** with family

Essence and the Interstitial Void

At the level of particles, quantum theory reveals that matter is governed by other laws and that it vibrates in the universe at multiple frequencies that can exist simultaneously in parallel worlds. These multiple worlds, or parallel universes, are interconnected by the gravitational effect. If the essence of all humans vibrated inside their interstitial void while coexisting with those of other people without entraining them, we would reproduce the quantum behavior of parallel worlds. It would be as though we were living in parallel worlds while coexisting, like matter does. Imagine that we have tuned our interstitial void on the frequency rate of our essence without entraining the other and without being controlled by the other. By choosing to diffuse our **essence** in our interstitial void, without influencing others, we increase our individual frequency rate to its maximum. We thereby amplify exponentially our personal signature and enable others to vibrate their profound nature without this causing conflict. When we function on the basis of the interstitial void and feel its beneficial effects, this leads to an absence of arrogance that always amazes me. Our goal is to vibrate our frequency rate without irritating others or creating competition.

In the universe, all probabilities coexist without canceling one another. International diplomacy would take a new turn if nations, instead of trying to align themselves with the most powerful ones, expressed their individuality while coexisting among themselves. Perhaps if we all tuned in on our own individual frequency rate and coexisted with those of others without entraining them, we would finally have peace on Earth!!!

By having the frequency rate of our essence vibrate in our interstitial void, we will be able to establish new relations with those close to us. Imagine that together we all emit our individual frequency rates without being mutually entrained by one another. We would then experience a rare and precious harmony. This reminds me of the concept of the tri-tone, a third sound that is created when we play two notes simultaneously and which is superimposed on the first notes. We can well imagine that in the future, communities will form spontaneously and naturally and in keeping with the individual frequency rate of each person. It will no longer be necessary to adhere to a cause or a religion in order to live in harmony. By remaining totally autonomous and vibrating our frequency, we will automatically be attracted to people who vibrate at a frequency in harmony with ours. This recognition will take place at the level of our genetic code.

In a couple, no partner is closer to vital energy than the other because consciousness vibrates at all levels at the same time and at different frequencies. We are all equal before the Source. No creature is closer to the Source than another. On our planet, we are all as close to the Earth as everyone else. There are no men or women who are more human. We are all equal when it comes to the vital force. In the universe, everything coexists. In nature, all creatures have the right to vital energy. There is no spiritual guide or experience that is closer to the Source than another, since everything that has been created exists simultaneously in Creation.

From now on, each person must become aware of her/his individual frequency. We initially embarked upon the path of consciousness so as to know ourselves better and then manifest what we want in our lives. This phase has come to an end.

The New Stage of Consciousness

We have reached a new phase where we can now celebrate our essence. We will vibrate our original color, our individual tonality. We will maintain a constant and natural vibrational state without resorting to external means. We are no longer seekers in search of the absolute, but people **who know their essential frequency.** Rather than believing that we have to either dance, play tennis, do yoga, travel, or some such activity, to contact our essence, we will understand that all these activities are of lesser importance and that the truly important thing is that we come back to our essence by using the power of intention at zero point.

Imagine that your feet are amputated after an accident. Would you be able to come back to your essence without dancing or playing tennis? Of course. Because contact with our essence is vibrational and has nothing to do with the form. By developing a stable relationship with our essence, we will manifest comfortable and healthy realities in our lives. Our professional, social and recreational activities will be entirely suitable because they will be in resonance with the frequency rate of our essence.

We will stop asking ourselves how to best take care of ourselves and gradually we will not only identify ourselves with our essence but we will embody it completely. The more we embody it, the more we will love ourselves because this will seem only natural to us. We will feel gratitude and respect for the energy that we are. In this state, it will be very difficult to make bad choices. Our decisions will always be in harmony with our essence and we will be able to stop worrying and being watchful. We will become the guardians of our frequencies. This watchfulness was just a sign of our separation. If we are in vibrational harmony with ourselves, we will appreciate our personal flavor. Our path will be fluid, constant and compassionate and it will resonate with the perfect tonality in each of us. It will be easy to participate in the movement of the Universe while preserving our individuality.

From now on each person must become conscious of her/his individual frequency. The Source is tending towards a new paradigm where each element will vibrate at its exact tonality. By vibrating at the proper frequency rate, we will automatically find ourselves at a place that suits us. There is a place in the cosmos that is suitable for Jehovah, builder and destroyer of worlds, but there is also one for St. Nicolas. The one does not exclude the other; they both exist at parallel frequencies. The world is a model of exploration, experimentation and friendly coexistence that includes a multitude of frequency rates.

When we see essence being manifested in a person, not only do we find that person beautiful and natural, but also **true and authentic.** We are attracted because this person radiates something harmonious. It is increasingly easier to acknowledge these people without judging them because we find them harmonious and natural. In fact, what disturbs us about other people are not their differences but the confusion of frequencies. We will soon be able to share moments of well-being with others without being influenced by their frequency.

Exercise 6
ESSENCE

You can use this exercise on yourself or as a therapeutic tool.

After answering the following questions and establishing your individual profile, you can choose from various tools such as prayer, meditation or DNA reprogramming to put the situation back at zero point.

If you choose DNA reprogramming, test all of the following questions to determine whether you obtain a yes or a no. Then, test to determine how many protocols you need to put these questions at zero point. (See DNA Demystified *and* DNA and the Quantum Choice.*)*

You can apply this exercise to a particular situation. If so, identify the situation.

TEST if the person knows that:

1. her/his priority need is not only physical or emotional but vibrational;
2. she/he needs to vibrate at the frequency rate of her/his essence to feel alive and healthy;
3. by living in harmony with her/his fundamental frequency, she/he will naturally manifest what she/he needs;
4. if she/he lives in harmony with her/his individual frequency, activities at the professional, social and recreational level will be perfectly suitable;
5. from now on she/he must:
  a. become aware of her/his individual frequency rate
  b. maintain a constant vibrational state
  c. be in contact with the vibration of her/his essence

After having completed this exercise, one of the students experienced a moment of bliss. She felt all her cells vibrate and saw her connection to the Source. She lived this experience while remaining well anchored in her body.

She was grateful to have gone through the previous steps, which had allowed her to decode her unconscious programs and learn to remain in her body. These steps were extremely important but they were not the final destination. Our final destination is awareness of the frequency rate of our essence.

JUSTIFYING OUR ESSENCE

When we choose to return to our original color and live safely within our own frequency, we do not need to justify ourselves. We can command to be safe when we live in harmony with the rhythm of our essence.

The same principle applies to people who, for example, have an essence that is of the calmer sort. The essence of a quiet person does not have to scintillate or be on constant display to feel good. This person could say, "I choose that my essential movement, which is peaceful, be amplified." We can harmonize our needs with our essence. Our essential need, apart from being physical or material, is also vibrational. We need to vibrate at the frequency rate of our essence and thus, feel alive. This is how we will become conscious of our worth. By harmonizing ourselves with our fundamental frequency, we will naturally manifest what we need without justifying ourselves. This will seem normal to us. For a long time we thought that we had to focus on our needs rather than on our essence.

On a beautiful white and blue winter's day, my son and I stopped at the top of a ski slope and took the time to communicate with a large fir. We tried to pick up the frequency rate of the tree's essence without analyzing it, simply by looking at the tree and trying to feel its essence. To do so, we imagined that we had a dial that served to amplify the frequency rate of our essences. By using an intention at zero point, we commanded that our essences be amplified to the maximum. After a few moments, we felt such peace and such affinity with the tree that the only word that came to our minds was **innocence**. We had just made our essences coexist with that of the tree and had felt a deep reverence and love for this forest entity. This experience created a bridge between our being and our essence. My son confided that, having spent a rather trying week at university, he finally felt that he had found himself again. There is no rivalry among the frequency rates of the various elements of nature. Even in the struggle between strongest and weakest, each holds on to the integrity of its frequency rate without questioning the identity of the other creatures.

In the French Forest of Brocéliande in Brittany, I was fortunate to come in contact with a 1000-year-old beech tree. Standing before this proud, beautiful and colossal being, I did not know what behavior to adopt. Deep within myself, I asked what I should do before such a powerful tree, a survivor of so many years of history radiating such splendor. The tree answered me through my intuitive self and told me to simply vibrate the frequency rate of my essence in its presence. So, I stood before the tree and requested the amplification of my essence in the presence of its essence. When I left the forest, I felt true and authentic with all my weaknesses and particularities. I had been moved by this sharing of our **essential** frequencies, which had required no adjustment on my part. There had been no exchange. I had experienced the harmony of our two tonalities forming a third tonality or tri-tone of love.

A tri-tone is a sound whose frequency is a multiple of the basic sound. During the workshop I led on the present book, I had the good fortune to share this experience with a student who felt a powerful amplification of her frequency rate after the exercise on essence. I am convinced that encounters based on the coexistence of our essence will allow us to strike up harmonious and melodious relationships in the future. It is exciting to imagine the type of relations we will establish during these **essential** encounters, in which two people each diffuse an authentic vibration, with these two vibrations together producing a tri-tone. We will finally experience true peace and friendship. We will become like superstrings, which vibrate simultaneously, side by side, at individual frequencies.

When we vibrate at the frequency rate of our essence, we no longer need to justify our existence. We have the right to exist, as do all the elements in nature, the animals and the stars. We may have gone through moments of doubt because we were not wanted by our parents, for example, but these wounds come from our environment, they are not who we are essentially. When we identify ourselves with our essence, we definitely have our place in the universe and a very specific individual tonality. When we finally connect with our essence and communicate with the people around us who are also in relation with their essence, we will create peace on Earth.

If everyone harmonized themselves with their essence and behaved as they pleased without justifying themselves, there would certainly be peace on Earth. This rectitude is that of essence. We would emit the appropriate frequency rate. We would stop borrowing inadequate quantum networks that are in harmony with someone else's frequency. Since everything coexists in the quantum physics model and vibrates at different frequency rates, the same would be true for us: we would coexist side by side with our different frequency rates. We are generally unaware of the frequency rates of those around us. We sometimes glimpse signs of their frequencies through their personalities, but we do not know these frequencies. Moreover, we can imagine what would happen if we all vibrated our unique essence. Instead of trying to convince others to vibrate at our frequency rate, we would be curious to see how the Source vibrates through the other. We would become responsible for our essence and we would no longer feel the need to justify ourselves. We could share our experiences without judging because we would know that each person is in her/his rightful place. We would finally be free of our codependent behavior, and this would happen effortlessly.

Our lives have become so efficient and programmed that we think we have to be likewise: high-performing and efficient. Life is not only about becoming perfect through self-growth, it is also about play. Once we learn to spend our lives manifesting all that we are, we will have a passion for the entity that we represent on Earth and will savor the expression of this entity's essence. We will blossom and be delighted with this frequency. We should not forget that our essence bears our signature in every world. It is the spark from the Source that shines from our Central Soul and we do not have to justify divine Creation! The path to the Source lies within.

Meditation on Our Essence

Each morning, we can meditate on our essence. We can be thankful and ask that it vibrate within. We can request its amplification and have more and more opportunities to live in the innocence of our unique frequency. We will feel the peaceful assertiveness that it leads to, regardless of the type of life we have lived, while holding on to our personality and character. We can imagine that if all the people we meet day after day did the same meditation and celebrated their uniqueness, life would appear much more interesting. The despair that is now crippling humanity would give way to hope and sovereignty.

"Thank you, dear Essence, for living in my body. Thank you for being aligned with the Source of all Creation. Thank you for being my connection to the Source of the universal Essence. Today I choose to bathe in the well-being and safety inherent to my Essence. And I command that the frequency rate of my Essence be amplified to its maximum today and forever."

Thank you, dear Essence, for living in my body.

I find it extraordinary that my essence can vibrate inside my physical body and not only within my etheric bodies.

Thank you for being aligned with the Source of all Creation.

Because of this experience, I have been able to live moments of satori, peaceful assertiveness and happiness.

Thank you for being my connection to the Source of the universal Essence.

I find it wonderful to know that, to communicate with the Source of Creation, I simply have to choose to vibrate my frequency rate using an intention at zero point.

Today I choose to bathe in the well-being and safety inherent to my Essence.

This ultimate experience of self will guarantee vibrational safety. When people vibrate at the frequency rate of their essence, they become impossible to manipulate.

And I command that the frequency rate of my Essence be amplified to its maximum today and forever.

As our frequency rate is amplified, it becomes easier to recognize it and to identify with our essence.

Visualization on Essence

1. Imagine that your essence fills your interstitial void completely and that the latter vibrates at your tonality, which has its very own flavor.

2. Observe your physical well-being when your physical body vibrates at the frequency rate of your essence.

3. Observe your emotional well-being when your emotional body vibrates at the frequency rate of your essence.

4. Observe your mental well-being when your mental body vibrates at the frequency rate of your essence.

5. Observe your spiritual well-being when your spiritual body vibrates at the frequency rate of your essence.

6. Imagine that your four bodies (physical, emotional, mental and spiritual) simultaneously vibrate at the frequency rate of your essence.

7. Command perfect health at all levels, in all your bodies, through the revitalization of your frequency rate.

8. Place your hands over your heart to feel your connection with your essence and the love that emanates from it.

CHAPTER 13
CIRCULAR HEALING

To have a good understanding of the basic principles of circular healing, we have to do away with the following statements once and for all: "I know I've gotten over that phase of my life"; "When will I ever get over emotions such as abuse, the fear of men or women, the fear of driving a car, etc."; "I don't understand why I'm still working on my childhood, my marriage, etc., because I thought I had dealt with all that." Any comprehensive process follows a circular path. If we want to be realistic, it is important to understand that healing is a state and not a destination. It progresses like a spiral rather than flying straight like an arrow to its target. It is a process of coexistence and of self-knowledge. It allows us to discover emotional, spiritual or mental subtleties that used to escape us. It reveals details about ourselves that allow us to live more fully in our body.

One of the beliefs hindering this process is the one where we think that we can make a wound disappear. Every wound leaves marks and makes us vulnerable. Therefore, we must change our relationship to the wound. We recognize good therapists not by the fact that they no longer go through emotional crises, but rather by the fact that they are able to remain present to themselves in moments of crisis. Aspiring to recover and enter a *perfect* world is a trap. We want to be able to recognize our wounds *while* remaining gently present to ourselves.

If we have the intention to consciously understand and experience what troubles us, we are certain to succeed because this attitude will get things moving. What was inside us that was blocked, rigid and tense will begin to move. According to certain theories, our unresolved and unconscious emotional wounds have the power to trigger illness. In my opinion, it is the tension resulting from the static nature of the emotional charge that makes us ill, rather than the emotional wound as such. We can reduce the tension these wounds cause, but they are an integral part of our past. We can transform our relationship to our wounds, but we should not, however, aspire to erase our childhood or the emotional betrayals we experienced. The danger of emotional wounds lies in their static unconscious nature.

Imagine that circular healing is like a pie and that one of the pieces, for example, is a defense mechanism that keeps us in poverty because of our upbringing. Imagine as well that we have worked so hard on this aspect of ourselves that we have reached our limit. And yet, regardless of all our efforts, no major change has taken place within us. Instead of targeting this aspect again, like we would do if we were following a linear thought pattern, we could ask ourselves if we would like to devote some time to the other pieces of the pie.

Instead of identifying solely with the pieces that would provide us with a sense of achievement, it is more efficient to overcome our fear of tackling another less interesting piece, such as the fear of being poor or of never owning our own house, etc., using a non-judgmental approach. Then, when we are sick or tired, we will insist on having the frequency rates of health and regeneration continue to vibrate in the presence of our symptoms. This could cause new subtleties, minute coincidences, intuitive feelings, achievements and moments of love and pride to spring forth. We would quickly grasp that a single piece, whether brilliant or positive, whether it represents achievement or carries a negative and suffocating charge, is less dynamic than the whole pie. We will understand that a single piece is less solid than the complete pie and this will prompt us to explore the other pieces. We will become more fluid; it will be easier to adapt to change and we will achieve balance at zero point. This new journey will become an initiation where all our efforts will allow us to become familiar with new individual aspects of the whole picture of who we are. Gradually, we will face life in a circular way.

We will look at what life brings us as an experience to be lived instead of a fight to be won. If today it brings us fatigue, then we take a break. If we are too edgy to rest, then we allow restlessness to exist and simultaneously accept our resistance to stop. To do so, we will need to change our criteria: from now on, the important thing is not our capacity or incapacity to rest, but rather the acceptance of all that we are while remaining in *essential* movement.

Our liberation lies in our intention to occupy all of the pieces of the pie that make up who we are. There are parts of ourselves that were traumatized during our childhood and they are now completely stuck. They plunge us back again and again into an inexplicable paralysis. If we want to take a holistic view of our being in these moments of powerlessness and stress, we will inhabit more than one ray in our circle and simultaneously access the vibration of the movement outside this wound as well as the non movement of its pain. We will enter into contact with the magnetic charge of our powerlessness and we will choose the coexistence of movement. By becoming circular, we will encompass the pieces that are not moving and will achieve conscious motion. When we are polarized, some parts of ourselves are in motion and others are completely stuck. By allowing all these parts to exist, we will acquire a new velocity that will express itself through circular movement. This new approach will allow us to transform our program and attract new experiences into our life.

When we only concentrate on one part of the circle instead of combining the elements of velocity with the elements that are stagnant, we obtain polarized and temporary results. Instead of wanting to change difficult aspects of our personalities, we can let them be and, by accepting them, come to terms with and explore their emotional depths. We can feel confused while effectively pursuing our goals and formulating effective intentions.

To reach a circular process, we have to accept all that we are. It is true that, to reclaim power, certain steps are prescribed, but since the pilgrim's path is not linear, who can say how the journey must proceed? In addition, it is crucial to remember that the goal of self-knowledge is self-love and not perfection. If we continue to be self-accepting and remain compassionate, we will definitely get convincing results. We have to stop thinking that we have to heal, improve, force ourselves to do things and proceed according to a specific religious code. The principles of the great religions and the great philosophies have had ample time to prove their effectiveness and have not yet succeeded in saving anyone from themselves.

The new quantum thinking that advocates chaos, multidimensionality and multiple possibilities is governed by the intention and choice of the observer. If we believe that our process will be long, it will be. If we wish for a quantum leap, we have to adhere to the principle of coexistence, self-love and love of our lives. I propose that you spend the next two weeks telling yourself that you love your life, the journey of your healing process and the part of you which is going through this stage of your healing. Rather than trying to develop higher consciousness, we could *choose* to accept wholeheartedly the process we are going through now *even if* we do not know how.

All the events in our lives are parts of the process and we cannot pretend they are not there. We need to choose to love ourselves despite our harmful self-criticism. When we decide to remain entirely in the here-and-now, without giving precedence to one aspect of ourselves over another, we can gain a more circular understanding that will help us stop being self-judgmental. We will have access to the power of our darkness and we will have a growing conviction that our intentions at zero point are truly effective. We will finally be able to stop looking for salvation and concentrate on being present to ourselves and our lives. To change frequencies and establish a new way of vibrating, we have to adopt a new way of thinking that exists within the state of love at zero point and includes light and darkness.

If we have been trying for years to improve our behavior and heal our wounds, we know the importance of comprehensive and holistic integration. By separating spirituality from our other selves, it may lead us away from our emotional wounds. The emotional process could exclude the transcendent aspect of spirituality and tidiness of mind. Mental reprogramming has a tendency to disregard our emotional resistance. Physical training is sometimes accompanied by a negation of emotions. In including all the pieces of the circle, we need to include the ones that have nothing to do with our wounds. According to circular healing, the wound is a just another piece of the pie.

Our identity must also follow the same path. Our wounds do not reflect all that we are and are not to become the definition of our evolution, nor our destination. When individuals endlessly speak about their wounds, it tends to hide their true identity. It is impossible to know what type of person they are. We identify ourselves with our wounds rather than with our true essence. A friend once told me that, during a conflict with his spouse, he had to clearly affirm that his being was not limited to what she saw in him, that is to say, the wounds that got on her nerves; he had to do this in order to continue to be true to himself in the relationship. Our essence is circular, but all possibilities coexist within it. We can have preferences, but all parts of ourselves need to coexist. Can anyone argue that there is a point on the circumference that is closer to the center of the circle than another?

Inner work and self-growth are so much more than just trials. They are meant to bring us closer to and radiate our essential frequencies. If we remain **circular**, we will be able to be proud and self-contained and we will be able to circulate our own energy within our own circle without creating an imbalance. When going through an expansion–professional or otherwise–we have to expect this growth to influence everything that we are, our dark and light parts alike. Thus, our difficulty to establish boundaries might be accentuated. Our childhood and transgenerational wounds will also be more pronounced. This is why it is imperative to remain circular and at zero point, no matter what is happening in our lives, so as to allow positive and negative experiences to coexist.

Exercise 7
INCLUDING OUR RESISTANCE

You can use this exercise on yourself or as a therapeutic tool.

After establishing your individual profile, you can choose from various tools such as prayer, meditation or DNA reprogramming to put the situation back at zero point.

If you choose DNA reprogramming, test to determine how many protocols you need to put these questions at zero point. (See DNA Demystified *and* DNA and the Quantum Choice.*)*

1. IDENTIFY the stagnant situation.

The following is an exercise that we use to help us manifest what we want. It consists of writing down what we want. For example, a woman who suffers from insomnia when she sleeps away from home could write the following phrases:

I can sleep peacefully away from home,

BUT I am bothered by foreign vibrations.

BUT I am afraid of being disturbed in my sleep.

BUT I am afraid of waking up with a sore back.

BUT I am afraid of not sleeping well and waking up with a headache.

BUT I am afraid of suffering from insomnia and feeling isolated in an unknown environment.

And so on until she can write: I can sleep peacefully away from home, without there being any BUTs that follow.

2. REDO the exercise and include the formulation "while" to put the intention at zero point.

I sleep peacefully away from home,

EVEN IF I am bothered by foreign vibrations or *WHILE ACCEPTING* my sensitivity to foreign vibrations.

EVEN IF I am afraid or *WHILE ACCEPTING* my fear of being disturbed in my sleep.

EVEN IF I am afraid or *WHILE ACCEPTING* my fear of waking up with a sore back.

EVEN IF I am afraid or *WHILE ACCEPTING* my fear of not sleeping well and waking up with a headache.

EVEN IF I am afraid or *WHILE ACCEPTING* my fear of suffering from insomnia and feeling isolated in an unknown environment.

Chapter 14
Circular Movement in a Stagnant Situation

Stagnation and Movement

Ascension is not a state but a vibrational experience. It is a process by which the cells of our body begin to vibrate in resonance with the frequency rate of our essence. Therefore, ascension means movement. Reality is not static: its behavior changes, depending on the observer, and there are parallel worlds with vibrational frequencies that differ from those of our own world. The more I understand these concepts, the more I realize that everything is a question of vibrational frequency. This applies not only to matter but also to events. Since quantum physics has proven that there are parallel realities and that they vibrate at different rates, we could choose to align our projects with a reality whose frequency rate is in resonance with that of our essence.

According to the principles of quantum physics, each observable event first belonged to a series of possible choices. It was in accordance to the observer's decision that one of these possibilities was manifested. All other possibilities–those that were not chosen–continue to exist in parallel worlds. Once the decision has been made, we embark upon an endless sequence of cause and effect. Thus reality can be defined as a series of possibilities that vibrate at different frequency rates. Each potential reality behaves like a wave. All realities coexist and emit different frequency rates on parallel paths. These are potential parallel realities. By changing quantum paths, we will set our project back in motion.

Essence automatically seeks a form of expression that is in resonance with its frequency rate. By choosing a new possibility based on our essence rather than on our unconscious programs, we will be able to place the situation back within another network of probabilities. We will be able to celebrate the pleasure of being in movement and we will be free of some foreseen, restricted future. Our essential movement will be our success. Time and space will become flexible. From then on, the road traveled will be less important than the journey itself.

Vibrational Rate of a Situation

A situation that does not evolve is not necessarily blocked. It may be on the wrong quantum path. In quantum physics, each situation is considered to have lots of possibilities and matter is defined by probabilities based on the number of times it has behaved in any given fashion. We can imagine each of these possibilities as being located on a parallel path.

When faced with a disappointing situation, we should learn to view our distress from a vibrational perspective. When our expectations have fallen short, it is possible that the cause of our disappointment may be of a vibrational nature. Instead of a blockage, it would be more appropriate to speak of stagnation, or the non-resonance of frequency rates. This means that our project is not on the right quantum path. Through the power of intention, we can redirect it to a path that is in harmony with the frequency rate of our essence and set it back in motion. Rather than voicing the intention that the situation begin moving again, our intention needs to be that the situation become fluid, and be repositioned on a quantum path that is in harmony with our essence. When we find ourselves on the wrong quantum path, we may very well expect stagnation.

Take, for example, a student named Theresa, whose love life had been non-existent for several years. After having done a lot of decoding and reprogramming, she felt her blockages dissolve; however, she was still not seeing any tangible results in her life. We then wondered if the situation might be stagnant. During the workshop, I drew a ray representing Theresa's frequency rate, and then I drew a second line on a different wave to illustrate the difference between the frequency rate of Theresa's essence and that of her desire to have a love relationship. It was as though she imagined her future spouse on a frequency that was calm and green, for example, while her own intentions vibrated at a red frequency. These are two completely different rhythms. Theresa must therefore place herself at zero point on another path in resonance with her frequency. If she repositions herself, doors will open without any effort on her part. Otherwise, all the energy she puts in building a love relationship will come to nothing because of the conflicting frequencies. It is like a sword slashing through water. Nothing moves.

We do not need to change or to become perfect to choose another latent vibrational possibility. We can have stagnation coexist with movement. We can choose a new path by using the power of the frustration caused by the former decision. Our new movement will not be polarized but will take into account everything that exists.

Take Peter, for example, one of my clients who was to leave on a cruise with his parents, his two sisters and their husbands. This trip had been his youngest sister's idea. She wanted to take their parents to Europe before they were too old. She had planned a two-week cruise during which they would visit a dozen European cities. After assessing his needs, Peter realized that he had gotten himself involved in his sister's venture, but that it really did not correspond to his current needs. I therefore encouraged him to express his feelings to his mother, who received it very well. In the end, she was relieved that he was not going because she needed someone to watch over her house.

Overnight, Peter found himself with a nice fat bank account, since he no longer had to pay the huge expenses related to this trip. So, he decided to invest in a condo that he would rent out to build up retirement funds. In no time flat, he found the ideal condo in a building he had been interested in for several months. This came as a real surprise to the owner who had not expected to sell so quickly. The bank approved the mortgage in a few minutes and the transaction took place in record time.

I had just witnessed an extraordinary change of quantum path. Peter was initially on a path that vibrated at his sister's frequency. When he identified his needs and changed paths, doors immediately opened for him. His lack of enthusiasm for this trip was equaled by his excitement at buying a condo.

When we find ourselves on a quantum path that vibrates at a frequency rate that is in harmony with our essence, we immediately start experiencing the beneficial effects derived from the power of this vibrational resonance. Our decisions and our actions become effective, we have a better grasp of the issues and the moments of synchronicity multiply. The current flows swiftly and with fluidity.

Vibrational Safety

One of my clients–we will call her Olivia–was preparing to move but felt that she and her husband were curbing each other's momentum. Her husband was getting ready calmly and without haste. When she found a house that appealed to her, her husband did not like it. If he wanted another house, she raised an objection. Together we looked at whether she could command a suitable place that would be aligned with her frequency rate and that of her husband. When we radiate our own frequency rate in the presence of another person, it is sometimes difficult to feel that we exist or that there can be a win-win situation.

Eventually, Olivia *stated the intention* of finding a place that was appropriate for her frequency rate, as well as for the frequency rates of her daughter and her spouse, even though she was feeling resentful. Once this step was taken, she had to define the term **appropriate**. What is appropriate for one person is not necessarily appropriate for another. Olivia and her family therefore had to express their preferences by first drawing up a list of their personal needs. She listed the following elements: nature, peace, ease and safety. **Safety** was very much her **priority need**. Interestingly, the type of safety had changed from what it had been at the time when she and her husband had bought their current house. Today, Olivia is looking for a feeling of **vibrational safety**. During our discussions, Olivia realized that she felt isolated in her neighborhood, not socially but vibrationally.

I sincerely believe that the communities of the future will be based on vibrational genetic recognition. We will be attracted by the frequency rate of certain people. Olivia sensed that if she felt vibrationally safe, she would not be afraid of movement. She even felt that she should begin consciously manifesting what she wanted. Instead of artificially maintaining a project at a tolerable vibrational level, she could decide as of today that her next home would correspond to her frequency rate.

For her, manifesting events that were in resonance with her essence had significant impact on her daily life. She understood that this type of manifestation was worthy of the great spiritual masters. This vibrational safety filled her with a peaceful feeling of power and balance. She no longer feared that her husband would **win**. Of course we could have already expected some affinities, since they had been married several years and had experienced many ordeals together. Nonetheless, vibrational resonances bring forth a different sort of affinity. Therefore, a project manifested in harmony with the frequency rate of the essence of one person–as long as it is not based on default programs–will automatically be comfortable for the other.

The following exercise will help you get your project back on a quantum path that is in resonance with your frequency rate. You must first identify a situation that does not seem to be moving forward and then state the intention to place it along the appropriate vibrational ray. We should not be afraid to change paths! We should dare to get our lives moving again by taking our needs into account. Keep in mind that we are not the ones who need to change, but rather it is our intentions that must align themselves with the frequency rate of our essence!

Exercise 8
STARTING THE CIRCULAR MOVEMENT IN A STAGNANT SITUATION

You can use this exercise on yourself or as a therapeutic tool.

After establishing your individual profile by answering the following questions, you can choose from various tools such as prayer, meditation or DNA reprogramming.

If you choose DNA reprogramming, test all of the following questions to determine whether you obtain a yes or a no. Then, test to determine how many protocols you need to put these questions at zero point. (See DNA Demystified *and* DNA and the Quantum Choice.*)*

1. TEST if the person knows how to get the situation moving again.
2. TEST if the person is enthusiastic about this project.
3. TEST if the person is afraid of getting worked up or of getting involved in something too big if this project were to get moving again and were successful.
4. TEST if the person can embark on this project.
5. TEST if the person can assume complete responsibility for this project.
6. TEST if the person can embark on this project at her/his frequency rate even if she/he:
   - **a.** is afraid of failing;
   - **b.** does not feel equal to the task;
   - **c.** has doubts;
   - **d.** feels another emotion (identify the emotion).
7. TEST if the person is absolutely convinced of meeting her/his needs if this project becomes reality.
8. TEST if the person understands how to have movement coexist with stagnation.
9. TEST if the person accepts to have movement coexist with stagnation.

10. TEST if the person knows where her/his doubts come from. If NO, using intuitive means, test to determine the circumstances behind the doubts and the age of the person at the time.
11. TEST if the person can affirm that she/he chooses a blessed future in regards with this project.
12. TEST if the person's fear can coexist with the expansion of this project.
13. TEST if this project can align itself with the frequency rate of her/his essence.
14. TEST if this project brings vibrational safety to the person.

Aligning a Project with the Wave of Our Essence

We often hear that suffering is our greatest teacher. In fact, suffering tells us that we are on the wrong quantum path. Finding ourselves on a path that is not in harmony with the frequency rate of our essence is so uncomfortable that this forces us to change paths.

As we saw in Chapter 12, if we find ourselves in a stagnant situation, that is to say in a situation that is on a wrong quantum path, this is because the frequency rate of our essence is too dim in regards with this situation. If we are not in resonance with our individual wave, we will not automatically respond by choosing a quantum path in harmony with our frequency rate.

Take Theresa for example, who was mentioned at the beginning of this chapter. Since she did not identify herself entirely with the frequency rate of her essence, Theresa was always choosing a quantum reality that did not correspond to her true nature. As a child, she had started to identify herself with a religious model of saintliness and chastity. She believed that she existed only to save others. In reality, her essence does not correspond to this model, which vibrates at a very specific frequency rate and which is different from this religious image of herself. The frequency rate of her essence, beyond the influence of her family and environment, is not that of a saint. Yet, instead of vibrating the frequency rate of her essence, she vibrates that of a saint. She therefore only succeeds in building platonic relationships because, when she vibrates the frequency rate of this energy, she is on the path of saintliness and chastity, which does not correspond to her deep nature. On the other hand, if her essence were that of a mystic, she would easily gather momentum on this path because the two tonalities would be compatible.

When we do not know how to resolve a situation that has become inappropriate or stagnant, we must ask ourselves if we are experiencing a perfectionist trance that excludes our personal needs and takes no account of our essential nature. Take the time to describe in detail this perfect scenario so that we can imagine ourselves at our best, all our needs having been met.

Take Monica, for example, another client who is a massage therapist. After having imagined a perfect scenario by listing her different feelings, we noticed that what she enjoyed the most as a massage therapist was not giving a massage as such, but rather her interaction with her clients as well as the sales aspect of her work. When she sees herself as a businesswoman, Monica feels very much alive, sparkling. Her priority need is to feel that she is moving and exuberant. Too often, we tend to waste time worrying about the particulars that lie behind our well-being rather than imagining what we would feel in this state. We are preoccupied with the form/effect rather than the state/cause. And yet the form keeps us on the wrong quantum path whereas the state defined by intent can enable us to access all latent possibilities. Since Monica **needs** exuberance, she chose to feel this way, trusting that its expression would come.

The emotional portrait of our needs is deeply associated with our essence. By imagining herself at the head of a health center (as an example of a desired form) she could feel her exuberance (the state), and Monica got a glimpse of the flavor of her essence: sparkling and exuberant. She then decided to manifest her essence in her daily life instead of restricting it to her professional life. By vibrating the frequency rate of her essence, she naturally expressed her personal flavor, which is exuberant. If we are not in contact with who we really are and with our **essential** identity, we continually try to advance along a quantum path that is not suited to our personality or our essence. Our energy cannot be effective. **Among all the quantum possibilities that exist, there are realities that vibrate at the same frequency rate as our essence. It is in choosing these realities that we can be truly happy.**

I advised Monica, who was wondering how she could manifest exuberance in her life, to throw this idea away, because, if she maintains the magnetic frequency rate of her exuberant essence and vibrates it, she will definitely attract the circumstances that will lead her to success even if the road she takes is sinuous. No longer wondering how to do it, she only has to call forth her essence. The constant element of her journey will no longer be a fictional straight line but rather the consistency of the frequency rate of her essence, which will permeate all her undertakings as well as the situations she will attract.

What is most surprising is that we feel afraid and insecure in the face of our essence. We are afraid that in expressing our deep nature and becoming as exuberant as Monica's essential nature, we will fall ill or lack something. Disconnected from our essence, we have become afraid of our uniqueness. We have been morally conditioned to believe that we cannot become our own life experts.

Numerous emotional techniques attempt to teach us how to live by expressing our individuality. Several of them are very interesting, but I believe that the time has come to let go of this way of doing things. The amplification of energy caused by planetary changes forces us to choose a quicker and more powerful path. For my part, I choose the path of intention. Instead of trying to find out who we are, we can command the maximum amplification of the frequency rate of our essence inside ourselves. We can command the amplification of this extraordinary vibration which is ours. Day after day, repeating this command, we will gradually start to feel this frequency rate vibrate within ourselves. Our body will begin to change because our biology is perfectly adapted to our individual frequency rate. Our body will be revitalized by our essence. We will feel greater vitality and will become aware of the power of this energy, which is a natural part of each one of us. We will radiate energy and our vibrations will be healthy. This return to the frequency rate of our essence will delight our bones and organs, and our breathing will become deeper.

Our entire body will be able to react according to the vibration of the frequency rate of our essence. It is clear that this attitude will automatically allow us to take care of our personal needs with authenticity because we will be the natural champions of our essence.

Take Marianne, for example, who wondered whether she should move or not. Faced with so much indecision, she decided to first command the amplification of her essence along with her fears and doubts. It is obvious that at first we will not be able to constantly maintain our frequency, but zero point is not that demanding. Keeping discouragement at point zero is the key. Getting back to her essence day after day and commanding that her reality align itself with her frequency rate, Marianne commanded that she be able to clearly see the right decision aligned with her frequency *even if* she did not know how. She could also have commanded vibrational safety *while* being undecided. The same laws that govern intentions at zero point are applied; the only difference lies in the focus, which has become the frequency rate of our essence.

From now on, our main project is to live at the frequency rate of our essence. We must observe how our essence manifests itself in our life. Our essence will become the mirror of our frequency rate and we will be surprised to see the form it takes. We will reassess our life in terms of our essence. All our possessions, all our actions need to be reviewed on the basis of their authenticity and their vibrational resonance. It will then be increasingly easier to successfully materialize our projects while encountering fewer and fewer obstacles.

By allowing ourselves to command, among a multitude of possibilities those that offer us vibrational safety, we will create lives filled with passion and ease. We will become confident and reflect the peaceful assertiveness inherent to our essence. We would benefit from fostering compassion to ourselves even if we come up against obstacles and are not always able to maintain the frequency rate of our essence in daily life or at work. We should be satisfied with our progress even if we still sometimes need therapeutic support. Becoming aware of our essence is a process of self-love in which self-criticism and perfectionism have to be positioned at zero point. It is normal to resist change and our feeling of discontent can coexist with our contentment. Our path bears our personal signature and we are its benevolent guardians. Even if we deem it long and arduous, we will certainly reach our destination because our destiny is to come back to our essential selves.

Exercice 9
ALIGNING A PROJECT WITH THE FREQUENCY RATE OF OUR ESSENCE

You can use this exercise on yourself or as a therapeutic tool.

After answering the following questions and establishing your individual profile, you can choose from various tools such as prayer, meditation or DNA reprogramming to put the situation back at zero point.

If you choose DNA reprogramming, test all of the following questions to determine whether you obtain a yes or a no. Then, test to determine how many protocols you need to put these questions at zero point. (See DNA Demystified *and* DNA and the Quantum Choice.*)*

TEST if the person:

1. knows how to align this project with a frequency rate that is in resonance with her/his essence;
2. knows that to start the movement, she/he must tune in on the frequency rate of her/his essence;
3. knows that to start the movement, she/he must be able to choose, from all the possibilities that coexist, a quantum path that vibrates in resonance with her/his essence;
4. is in contact with her/his essence, regardless of resistance;
5. can have wounds coexist with her/his essence;
6. identifies with her/his wounds rather than with her/his essence;
7. can intervene in the stagnant situation by voicing the following commands:
    - **a.** If this project is intended for me, I command that it unfold easily.
    - **b.** If this project is intended for me, I ask for a sign.
    - **c.** If this project is intended for me, I ask that it be fun, etc.
    - **d.** If this project is intended for me, I command vibrational safety.
8. can obey these signs;

9. can give herself/himself time and compassion regardless of any resistance to change;
10. can remain attentive without falling into self-criticism or overachievement;
11. can say YES to the resistance and to the project at the same time. For example, "I command that my project begin safely while being afraid."
12. knows that she/he is preparing to carry out the paramount project of her/his life, which is to create a life aligned to the frequency rate of her/his essence.

Our Choices Influence Reality

Imagine that reality is a series of parallel lines of various colors on a blackboard. Each time we draw chalk along the red line, it becomes darker and darker. This red line is surrounded by other lines. We could draw our chalk over a pale yellow line and it would gradually become darker. Our choices have about the same effect. By making the same choices over and over again, we provide texture, color and depth to one line rather than another. Feedback, or the lack thereof, are always related to an appropriate or inappropriate vibrational choice. LET'S ALIGN OURSELVES WITH OUR ESSENCE!

If our project is stagnant, this is because we are on a line that does not suit our essential frequency rate. We therefore have to go back and identify our needs in regards to this project. Perhaps we have chosen a quantum path that is not in resonance with our essence by basing our decision on outside factors, rather than on our essence. It is easy to determine if we are on a wrong quantum path because, if we are, the tonality is distorted and the excitement and enthusiasm have disappeared. By setting ourselves back at an appropriate frequency rate, there is a good chance that we will experience unprecedented success. There is no need to fear change in this instance, because when frequency rates are in harmony, this creates great comfort and deep harmony. A situation that is in harmony with our frequency rate will automatically suit us.

Take for example a woman who would like to take her professional life in a new direction. She updates her resume. She clarifies her intentions, contacts potential employers and speaks with people who work in her field. She lines up interviews, mentally prepares herself, undergoes therapy, etc. All without any results. She would have good reason to ask herself if she is not on the wrong quantum path. When we have done everything to no avail, we need to look at whether the frequency rate of the situation is aligned with that of our essence.

To set a stagnant situation back in motion, we must understand that everything vibrates and that everything that happens to us depends on the frequency rate chosen. Having understood this, we can then choose the frequency rate we wish to tune in to. Then we can be at the right place, at the right time and in harmony with our actions. Our life will become more harmonious and we will be more effective without feeling stressed. When I speak of harmony, I do not mean that we will no longer experience arguments, but that there will be a certain vibrational resonance among ourselves. There will be movement and our projects will require much less effort.

If our life appears stagnant, we can give it new impetus instead of criticizing and analyzing the causes of this stagnation. This will keep us centered instead of scattered outside ourselves. We will try to amplify our frequency rate. We will instill movement in our life based on our essence instead of wasting our energy complaining. Tired of expecting miraculous events that will bring a much needed change in our lives, we will become the precursors of the rhythm in which our essence can blossom. By maintaining an adequate frequency rate, we will automatically be able to foster momentum in our lives. Staying in contact with our essence eliminates energy loss. All our energy is finally used at the right frequency and is not lost, as is the case when we try to live in the physical world without taking our essence into account. Imagine the power of the awareness that comes from decoding the path back to the Source, through the maze of the earthly dimension's limitations. To think that the frequency rate of our essence is the way to bridge this gap!

It is possible to maintain a specific frequency rate even if things are not going as planned. We can amplify the vibrational rate of our essence by making the most of the negative emotional charge overwhelming us at the time. All our experiences can be reframed to amplify the frequency rate of our essence. Sickness, happiness or distress: all our experiences can serve to command this amplification. We have to use our emotions as the driving force of our commands. *Let us command* situations that vibrate in harmony with our essence *even if* we do not know how.

In the end, if you are wondering why we should realign a project and become so involved in manifesting our reality, the answer is simple: are we happy where we are, doing what we are doing? If the answer is positive, it is because our situation is aligned with our essence. If it is negative, dare to make a new choice by specifying the frequency rate on which you wish to align your life. Do not be afraid of commanding vibrational safety. When feeling threatened, we must first ask ourselves if we have made our choices in resonance with our essence. In such an instance, we can use our fear as the driving force of our command and ask for vibrational safety. We can command that all our future decisions regarding all possible and probable possibilities be aligned with the frequency rate of our essence. By doing so, we will avoid vibrational distortions and ensure our vibrational safety. We are in danger when our unconscious decisions keep positioning us on the wrong vibrational path. If, for example, our spouse is going through an intense period and vibrates at a rate that is distinct from ours, we will not need to defend ourselves to feel safe, if we have stated the intent to be vibrationally secure.

When I have voiced this intention in all types of situations, I have been very impressed by its immediate results. Some projects immediately fell through whereas others that provided me with vibrational safety developed rapidly. I had to let go and accept to live this experience as an adventure bringing me back to my essence and, ultimately, to its source: the Central Soul. Keep in mind that any change at zero point that is aligned with our essential frequency rate leads to something better. We should not be afraid of the doors that will close because from now on we are moving at a high velocity. Our reality, thus managed by our essential frequency rate, will not tolerate any wave distortion.

It is at this point in our evolution that we begin to grasp the impact of the Central Soul, which gives birth to the frequency rate of our essence while seeking to manifest itself harmoniously in our terrestrial life. We can direct changes by seeking to activate our vibrational wave. This new phase provides us with the possibility of experiencing intentional manifestation. If we recreate our life based on our individual frequency rate, we will no longer be able to forget who we really are. Gradually, we will discover our splendor and our uniqueness.

Here is a little exercise that will help you break out of stagnation. In general, when we have identified the main cause of stagnation, things fall into place and begin to move again. First, answer the questions by yes or no. Your answers will give you a good idea of what has caused the stagnation. Once this cause has been identified, we will be able to proceed with the concept of coexistence.

Exercise 10
FORMULATING AN INTENTION BASED ON OUR ESSENCE

You can use this exercise on yourself or as a therapeutic tool.

After establishing your individual profile by answering the following questions, you can choose from various tools such as prayer, meditation or DNA reprogramming.

If you choose DNA reprogramming, test to determine how many protocols you need to put these questions at zero point. (See DNA Demystified *and* DNA and the Quantum Choice.*)*

1.New project aligned on the frequency rate of the person's essence and personal needs:

A. DRAW UP a list of the person's needs regarding a project that she/he wants to launch, aligned on the frequency rate of her/his essence. (See the List of Needs on page X). Before beginning, test to determine the number of needs. For example, there might be four

security
freedom
balance
abundance

B. IDENTIFY which of this person's needs is the priority need to be included in the intention. For example, it could be "professional abundance".

C. CREATE the first part of the intention for the appropriate vibrational ray *including* the **priority need**.
(The appropriate vibrational ray for our project is the one that vibrates in resonance with the frequency rate of our essence.)
For example: "I command professional abundance in resonance with the frequency rate of my essence..."

2. Resistance

A. CREATE the second part of the intention for the appropriate vibrational ray *including* the **worst fear**.

B. DRAW UP a list of all of the person's resistance regarding the stagnant project. Determine the number of resistance before starting (in this case 9). For example: If I had a better-paying job:

1. I would lose my freedom;
2. I would be working all the time;
3. I would have no more leisure time;
4. I would have no more time to devote to my family;
5. I would not be able to take care of my own needs;
6. I would not feel safe;
7. I would risk burnout;
8. the quality of my life would suffer;
9. I would lose my sense of balance.

C. IDENTIFY the person's greatest **resistance** by discussing the above answers. If, for example, the person is afraid of experiencing burnout, ask the person to say: "**I command professional abundance in <u>resonance</u> with the frequency rate of my essence** *while being afraid of experiencing burnout.*"

PART IV

THE ASCENSION PROCESS

Chapter 15
The Central Soul

The Central Soul is the vessel of essence and essence is its vibrational signature. Initially, the Source projected itself into different manifestations to gain self-awareness. The Source manifested distinct parts, each with its own unique vibrational signature: these are the Central Souls who were released into the cosmos so that they could manifest their vibration and become incarnate in all dimensions and in all worlds. This forms a network of interacting and intersecting probabilities which, throughout all these experiences, bears the Central Soul's unique signature. The information from all the incarnations that bear this same essential signature is available simultaneously at all times. In addition, our essential signature, which originates from the Central Soul, remains the same from one incarnation to the next and from one dimension to the next. Cosmic laws, like the law of free will, are applied throughout Creation. The Source did not create one, but many Central Souls who each received the gift of free will in order to explore the whole cosmos without limits. With an essence that is unique, a Central Soul expresses itself throughout Creation by exploring manifestations that are aligned with the frequency rate of this unique essential frequency. It acts like the head office of a large enterprise.

I recently had a spiritual revelation. I was seated quietly on a ski-lift when I felt energy reach me directly from my Central Soul–somewhere in the middle of the cosmos–through my hypothalamus. Not a small dose of energy, but rather a powerful creative current in complete harmony with my essence. I felt it spread throughout my body's interstitial void, where it amplified its frequency rate. There was no imbalance because the Central Soul is, in fact, the Source of our essence and encompasses all frequency rates. From its center, this essential frequency rate spreads out to all its manifestations and all its incarnations.

It is as though a source of my intrinsic energy was always readily available and continually vibrating at my own unique frequency rate. However, to properly understand this principle, we have to grasp the concept of the superstrings in quantum physics. At the quantum level matter is made up of strings and its nature is defined by the various frequencies at which these strings vibrate. It is no longer the chemical nature of an element that is important, but rather the vibrational frequency rate of its strings. **It is the frequency rate of matter that defines its individuality.**

I suddenly realized that the Central Soul, which is in direct contact with the Source of Creation, is a perfect relay and that I have access to it through my essence!!! It was as though I had just found my place in the universe. I had just glimpsed the superstring that I truly am. I finally had a vibrational identity.

At the beginning, I accepted this on faith, hoping it was real, but I was soon convinced of the truth of this realization. In the end, I developed a circular understanding that allowed me to reconnect myself to my Central Soul. The essence that defines me is the same in all my incarnations and this essence is the signature of my Central Soul.

I then saw that my allies (or spiritual guides) were in fact other rays of my Central Soul. Amazingly, these allies and I share the same essence. These other rays come to my assistance and bring me their support, doing so with a unique objective: that I identify myself more and more with our essence, because I share the same essence with these allies–they simply express different forms of my Central Soul.

Defining the Central Soul

When we are in a body, we have a soul that connects us to our Central Soul and each of our lives is a projection of the Central Soul into a dimension or a space/time environment. Free of all temporal definitions, the Central Soul is at the hub of all these incarnations, which unfold simultaneously along different rays. Central Souls are individualized parts of the Source and communicate with It directly.

Within each ray of the Central Soul, each life unfolds separately from the others. The information from these multiple rays is stored in the Central Soul. It is as though it were the central memory where all the experiences of these different lives were recorded.

The Central Soul chooses the types of incarnations that it wishes to experience. Our soul is what supervises our current life to ensure that it meets its objective and then it merges once again with the Central Soul. The soul has no other function as such.

To have a better understanding of this concept, I have created the following diagram, which represents a sort of sun and its rays. The center of the sun corresponds to our Central Soul and each ray represents one of our lives. This Central Soul is our unique signature and is associated with each one of our incarnations. All our lives spring from this Central Soul, which is in direct contact with the Source.

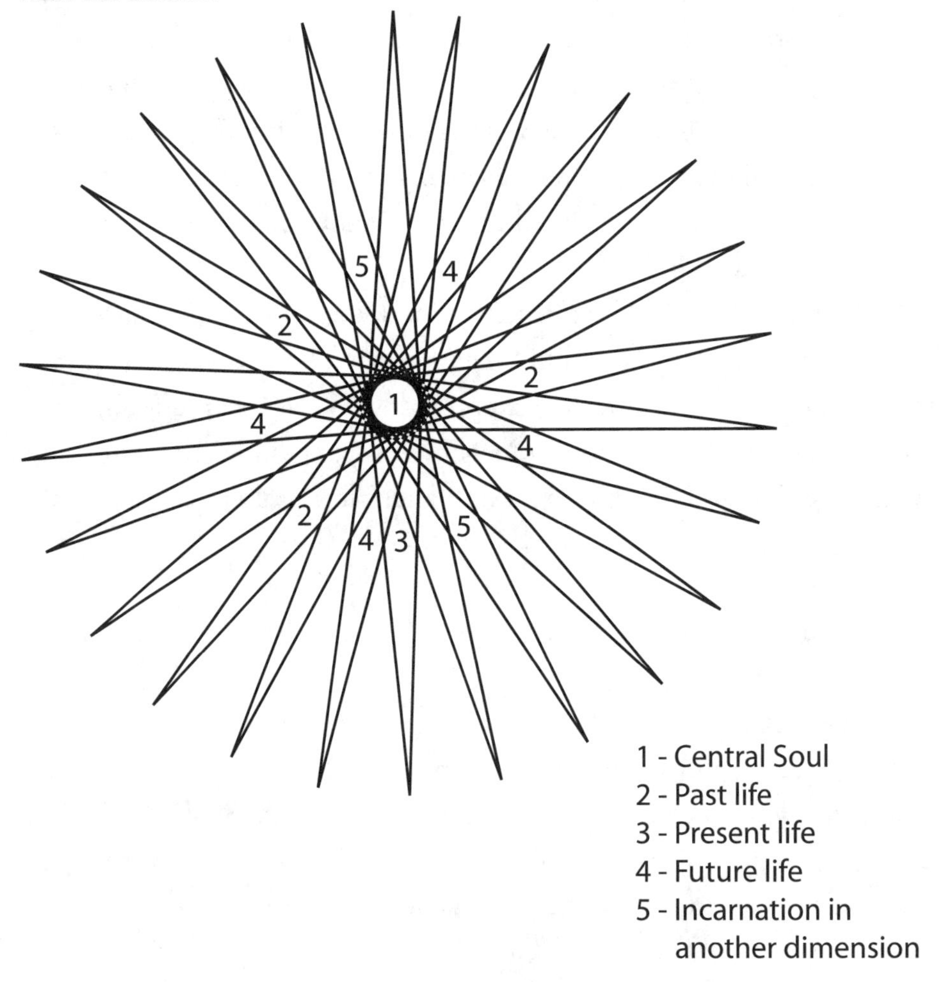

The ray that represents an immediate past life is not necessarily located next to the one that embodies our current life. Similarly, our current life is not necessarily the one that precedes our next life.

All the rays that surround the central sphere coexist, like matter does at the quantum level. These rays are not always completely straight. They can curve slightly and thus cross the beam of another incarnation. When this occurs, we say that the lines of time intersect. What is more, if we could look at this diagram through glasses that allow us to see things in the sixth dimension, we would see a sphere and a grid. Of course, if we looked at it through a window in the eleventh dimension, in which the strings are coiled upon themselves, we would have even more difficulty explaining the interaction among the different axes in a linear way, and our current position in respect to another life. Even if we perceive them as such, time and space are not linear. At any rate, what matters most is the relationship that exists among all these lives.

Imagine that each ray represents one of our incarnations regardless of the universe or period. These manifestations of ourselves exist in the past, the present or the future. They are embodiments of ourselves that may be found on Earth, in other worlds or in other spiritual dimensions. All our incarnations exist simultaneously and emerge from this central source. This multidimensional understanding of matter and of ourselves within Creation induces us to let go of the orderly image of the present that follows the past and precedes the future.

By reconnecting with the Central Soul and having all our incarnations coexist, we will rediscover our connection with the Source, who is in direct contact with the center (or Central Soul).

The Central Soul as a Network

Quantum physics has changed my way of thinking and influenced how I elaborate concepts. The image of the sun that I have previously used is still valid, but from now on, I view the "Source–Central Soul–embodied soul" model like a flexible grid or a network of possibilities made up of lines representing all our lives. The Central Soul's manifestation appears to me to be a grid system in which the space-time dimension is influenced by the choice of the observer. In thinking in terms of a network, I view all lives as fluctuating within this network. They become entangled, draw together, and then move apart. They are always moving and intersect at certain points. The points where they intersect are less important than the whole network. Indeed, the entire network vibrates under the command of the Central Soul and its unique frequency. This coexistence of all our incarnations and our link to the universal Whole define our place in the cosmos.

The Central Soul is no longer a specific point but a vibrational network that bears a very distinctive signature.

Each of our incarnations can exist simultaneously in this network. This means that at this very moment, we can access all that we have been and all that we will be. Creation makes no judgments. We are all that we have been and all that we will be. This discovery is very liberating. Nothing is more inspiring than seeing people radiate at zero point when they incorporate all aspects of their beings. Instead of identifying ourselves with a single incarnation or being limited by it, we will know that we are a true mosaic of diversified elements.

From now on we are able to appreciate all these parts. Self-love directed at the Central Soul, which is in direct communication with the Source, gives us the conviction that we are one with all that exists, beginning with ourselves.

This revelation filled me with a deep feeling of peace, order and harmony. My metaphysical evolution, inspired by quantum principles, is prompting me to put an end to the duality that exists among my different incarnations and to put them at zero point. I want to have access to the information inherent to all that I am. I want to integrate all parts of my being and enable them to coexist consciously. I want to feel connected to my Central Soul and be connected to the Source through the link that unites the Divine Source to my Central Soul. I can see that the return to the Source that is so strongly advocated by spiritual masters is in fact the awareness of our coexistence with It.

The Allies and Our DNA

At one point, before I started to work with DNA, I decided, since I did not clearly hear my allies, to take the bull by the horns and co-create with the Source within in order to obtain what I needed. In other words, I stopped waiting for my allies to save me and tell me what I should do regarding my health or other issues. With this action, I broke the cycle of spiritual wishful thinking. Without realizing it, I had just chosen to co-create what I needed and to regain sovereignty over my DNA.

Once I began my work on DNA, I heard my inner voice more clearly. Surprisingly, it was only later, after I had activated my 13 helixes, that I contacted my allies! They introduced themselves as vibrational allies that were linked to my 13 helixes.

Thus, our helixes are not only connected to our chakras, but also to our allies, co-creators, messengers and intergalactic relays. Since that moment, I no longer need to pray, search for my personal allies or my tribe, find where I fit in the big picture or establish contact with the universe. I know that the answer vibrates in the heart of each and every cell within the spirals of my DNA.

In addition, these allies of our DNA vibrate at the same frequency rate as our essence. In other words, these are other incarnations of our Central Soul. It is impossible to have an ally who is not in resonance with the frequency rate of our essence. Everything that we are is not expressed in a single incarnation. Several manifestations of our essence exist in other worlds and other spiritual dimensions. We can access the information relating to these other lives through the Central Soul network. We can even channel our own selves. We can send messages to our other incarnations and vice versa. We are interrelated through a single and unique frequency rate.

Of course, we can be inspired by masters who have ascended, angels or shamans, but none of these beings is closer to the Source than we ourselves are. They are governed by the same cosmic laws that we are. They can pass on to us a feeling of universal love or hope, but their path is the same as ours, although their essence is different. The time of the allies has finally come: these other selves that spring from a Central Soul that itself has emerged from the Source of Creation. Their sole objective is to bring us back to ourselves. There is nothing more powerful than this return to our essence, especially on Earth where we are all subject to the laws of duality and limited by time and space.

Communicating with the Central Soul

A woman wrote to me that she felt ill at ease because she found that her Central Soul did not communicate with her personally. She felt that we were going through very trying moments on Earth and that she needed leadership from her Central Soul. She added that she had tried to contact it through prayer and meditation and that she would request answers through dreams or visions. She lamented the fact that she did not receive answers and did not feel supported. I answered that I was certain that the information that came to us from the Central Soul was vibrational and that it was a wave that bore our signature. She answered, *"Thank you, thank you for reminding me. I have a terrible habit of expecting my Central Soul to speak to me in words and explain things to me."*

I was happy to have helped her discover the language of the Central Soul, but I remained perplexed. I knew that the essence that originates in the Central Soul is a wave and that we vibrate at this frequency. However, I did not understand why the Central Soul did not communicate with us better and vice versa. It took an accident and several long hours of waiting in a hospital corridor to finally get my answer.

I had a revelation regarding the Central Soul while waiting at the hospital. I was in pain and was wondering why I was not receiving help from the Central Soul. I was listening to sacred music on my walkman and waiting patiently while praying and meditating. Then I had an intuitive revelation.

I already knew that the Central Soul was connected to the Source of Creation and that all our lives in all dimensions sprang from the Central Soul. But at that moment in the hospital corridor, I understood that this was not a linear, but a quantum system. All lives are simultaneous and in quantum fluctuation. All the choices that we have not made continue to exist in parallel worlds. It was no longer a direct line from the Source to the Central Soul, and from the Central Soul to us, but rather a moving matrix. I understood that I had access to the <u>information</u> from all these possibilities that vibrated at the same frequency rate as my essence but that it was possible that the Central Soul did not know where I was!!! After all, what is a little hospital corridor among all these possibilities?

All these lives needed to be considered, from a quantum perspective, as being part of a network of simultaneous probabilities; the same was true for all the choices that I had not made in this life and which continued to exist in parallel worlds. I saw this virtual network, this matrix that vibrated and existed on all sorts of levels at the same time. In seeing the size of this network, I realized that if, on the one hand, we could receive an amplification of the frequency rate of our essence via the Central Soul, vibrational communication could also take place in the other direction.

I therefore contacted the vibrational signature of my essence by means of an intention and sent this wave to my Central Soul. I immediately received a wave of energy that was of great peaceful power. I felt peace, but above all great pleasure in knowing that I had been found. The nurse came to get us a few minutes later and we were finally able to see the doctor, who freed me of my cast while offering me the unprecedented possibility of a new cast that could be taken off at will like a ski boot. I had just been seen and helped by the doctor and the nurse but, above all, I knew that I had just received help from the Central Soul.

Now that I have reestablished contact from this incarnation in the third dimension, I know that I have just installed a new paradigm at the level of the quantum network of my essence, which originates from the Central Soul. I believe that when we succeed in giving life to a concept like this one within the Earth's density, it takes root and will develop like a solid and well-rooted tree.

In visualizing this matrix, I realized something else. I understood why prayer works so well when several people pray at the same time. There are so many possibilities in a quantum network, such as that of the Central Soul, that the focus of several people at a given point materializes a particular avenue, which facilitates a response in the Beyond. This focal point transcends multidimensionality and multitemporality. There is no longer a struggle among several opinions, different perceptions and viewpoints. There is only a love relationship, peaceful and confident assertiveness...

Since that particular moment, I have been experimenting with ways to maintain consistency with my essence. There is less and less conflict within and it is easier to occupy the center of my being. I know that I can receive an amplification of the frequency rate of my essence from the Central Soul, but I can also communicate clearly and feel a connection with It.

Exercise 11
RECONNECTING WITH THE CENTRAL SOUL

You can use this exercise on yourself or as a therapeutic tool.

After answering the following questions and establishing your individual profile, you can choose from various tools such as prayer, meditation or DNA reprogramming to put the situation back at zero point.

If you choose DNA reprogramming, test all of the following questions to determine whether you obtain a yes or a no. Then, test to determine how many protocols you need to put these questions at zero point. (See DNA Demystified *and* DNA and the Quantum Choice.*)*

TEST if, when reconnecting with her/his Central Soul, the person:

1. knows that the Central Soul is the Source of her/his essence and vibrates at exactly the same unique frequency rate;
2. knows that the Central Soul is the Source of her/his essence and is always available regardless of emotions and crises;
3. knows that her/his essence is born of the Source and is diffused to all its manifestations and incarnations through her/his Central Soul;
4. knows that essential energy comes directly from her/his Central Soul and comes through the hypothalamus;
5. is in contact with the creative current of the Central Soul, which is powerful and in harmony with her/his essence;
6. knows that the frequency rate of her/his essence can be amplified in her/his interstitial void;
7. knows that, like superstrings at the quantum level, her/his individual frequency rate makes her/him unique in Creation;
8. knows that she/he has her/his own vibrational identity;
9. knows that the Central Soul is in direct contact with the Source of Creation;

10. knows that the Central Soul is a perfect relay to the Source of Creation and that it is possible to access it through her/his essence;
11. can feel that she/he has found her/his place in the universe;
12. can feel that she/he has found her/his place in the universe without having to control or influence others;
13. knows that her/his essence is the same in all her/his incarnations;
14. knows that her/his allies are other rays of her/his Central Soul and have the same essence;
15. knows that identifying with her/his essence to reconnect to the Central Soul and the Source is one of the fundamental goals of existence;
16. knows that in choosing to express and feel her/his essence at zero point, regardless of the situation, she/he will lighten up, regenerate herself/himself and feel her/his body vibrate more and more;
17. can reconnect with feelings of:
    - **a.** hope
    - **b.** trust
    - **c.** certainty
    - **d.** conviction
    - **e.** faith in herself/himself
    - **f.** assurance
    - **g.** tranquility
18. knows that her/his essence is the vibrational signature of her/his Central Soul;
19. knows that she/he can contact her/his Central Soul from Earth by sending a vibrational message that bears the signature of her/his essence.

CHAPTER 16
CONSCIOUS ASCENSION AT ZERO POINT

DEFINING ASCENSION

The ultimate destiny of women and men is ascension. In the Christian religion, ascension is described as Christ arising above the heavens and disappearing from the sight of the disciples at the end of His mission on Earth. The Conscious vibration now available is quite different.

The great planetary activation that we are currently experiencing, and which is leading us towards a climax in 2012, indicates that we too shall experience ascension when we arrive at the end of our planetary mission. Ascension provides a broader perspective of our current reality. It is an acceleration towards multidimensional life that implies a change at the level of our current biology.

Ascension is a change in the vibrational rate of the body's cells, which begin to vibrate at the frequency rate of self-love and of our essence. The entire body begins to vibrate, which allows it to change worlds. We will be able to access parallel worlds without losing contact with the world in which we presently are. Quantum physics has demonstrated the existence of parallel worlds that exist simultaneously and vibrate at different frequencies. Conscious ascension will allow us to tune in to several frequency rates and access dimensions other than the current earthly dimensions. We will not depart from here and leave everything behind us, but we will have unlimited access to the various worlds that coexist with ours. Some of my students have already begun to experience this phenomenon. They wake up in the middle of the night and are no longer in their room or in this world. This experience, of course, seemed a bit traumatizing to them at first. But we must remember that these other worlds are governed by the same laws that govern ours. Here, like elsewhere, zero point, self-love and intentions are the primary tools. We will gradually spend more and more time in an ascended state and certain parts of our bodies will achieve this state faster than others.

Conscious Ascension

There are two ways to ascend. The first–simple ascension–is that achieved by Christ. Death is followed by a resurrection; the physical body disappears and becomes a body of light.

We can also ascend as a result of the harmonization of our essence and self-love. In conscious ascension, the entire body vibrates at the frequency rate of self-love and of our essence. When these two frequency rates **vibrate** simultaneously, they produce a third tonality, a tri-tone that is superimposed on the first two. This third vibration is that of ascension. It is the frequency rate of this new tri-tone that triggers the ascension process.

In this second case, ascension is conscious. The body of light adjusts to the physical body without bringing about its death. When we voice the intention to ascend consciously, our body begins to change and seeks new ways to come into line with this decision. To ascend, the body has to evolve. It must achieve vibrational autonomy. The process of ascension is 'gradual' and it is possible that certain parts of our body have already ascended. Some days we will appear to have gotten younger and other days we will see no difference. The more we identify ourselves with our essence, the lighter we will become. Our body will increasingly bear the signature of our essence. Gradually, we will see our body as the vehicle of our essence and we will revere the physical manifestation of our essential frequency rate. By constantly maintaining our intention that this frequency rate vibrate inside ourselves, we will feel a deep innocence that will allow our body to change frequency rates and align itself with our own unique frequency rate.

The more we experience this frequency rate, the more we will want to maintain it, and our experiences of well-being, self-love and ascension will automatically multiply. We will understand that illness is the result of a vibrational shortcoming and that our body, with its high percentage of void (±70%) is simply the vessel of this vibration. We will become the sacred vessel of the divine manifestation of essence, for which ascension is an innate right.

Ascension and Essence

Nowadays the whole world is expected to share the same emotions and aspire to the same lifestyle. If we could only conceptualize the scope of essence, we would use our free will with the sole purpose of maximizing the amplitude of the frequency rate of our essence. We would be free to consciously choose our reality in a universe of infinite abundance where we would all be vibrationally different from one another. The only real 'mistake' at this stage would be to consciously make a choice that is not in keeping with our essence. The greatest sin consists of censuring our essence. In my opinion, that is the original sin. Censuring our essence is worse than censuring ourselves. This does not mean that our life is not one of the manifestations of our essence.

Our essence is not present in this life only; it manifests itself in all our lives. When we censure our essence, it is not only this life that we censure, but also the expression of the Source. Through the ascension process, we will develop greater awareness of our essence. Without our essence, we would not be able to live because it is the vital energy that keeps us alive. In fact, without our essence, we would not be part of Creation. It is also through our essence that we will return to the Source.

The Source of Creation communicates with us through the Central Soul, in keeping with the cosmic order. This calls to mind the basic principle of alchemy. Our essence is an integral part of the Creator's essence. From the Whole comes Unity. The alchemist model presents Unity as being nourished by the Whole and nourishing the Whole. The frequency rate of our essence is part of the frequency rate of the Source and the Whole of the Source includes the frequency rate of our essence. The drop becomes part of the ocean, while remaining a drop. It is nourished by the ocean and simultaneously nourishes the ocean. According to Edgar Cayce, the Creator created us so that we could choose to return to the Source of our own free will. It is at this precise moment that the Cosmic Flux enters into an alchemical relationship with Its Creation. By identifying with our essence, we will nourish the Cosmic Flux and be nourished by It through an eternal exchange and coexistence at zero point. An extraordinary love story exists between the Central Soul and the Source. Contacting our essence and returning to the Source through the Central Soul is the path that leads to self-love and the limitless matrix of Universal Love. The relationship between the Central Soul and the Source is neither religious nor metaphysical: it is a quantum relationship. To understand this, we need to know that everything that exists is part of a vibrational network from which nothing is excluded. This is not only valid for our holy or spiritual side because our essence sets its signature on all aspects of ourselves, good or bad. The concept of zero point enables us to finally accept ourselves.

Ascension and Self-love

Ascension is the energetic manifestation of the frequency rate of self-love. When we love ourselves unconditionally by incorporating the best and the worst of ourselves, our cells vibrate faster and the energy of the photons increases, producing more light. To ascend, we must be able to accept all aspects of ourselves, because this light will reach into all aspects of our being!!! It is therefore important to accept everything about ourselves at zero point. We need to love our better aspects as well as our darker aspects, even if we don't like them. We are required to stay present to our anger, fear, disappointment and confusion that might be awakened from deep within when we go through this frequency activation. When we lack tolerance and feel dissatisfied with ourselves, the best approach is compassion. Even though we may be disoriented because of the activation of this frequency rate, we will stay present to ourselves through our compassion. To be truly authentic we do not have to be perfect; we simply need to not polarize ourselves in self-judgment. We can love ourselves while criticizing ourselves.

All religions preach self-love. If there is a universal code, this is it. To ascend, we must access this code. The most functional intention is without a doubt the following, "*I choose* to love myself *while* not loving myself!" However, self-love and self-knowledge go hand in hand. Indeed, the more I know my essence, the more I love myself. I recognize myself in my essence and I learn to appreciate myself. When I identify myself with my essence, I discover a new identity. I am no longer the therapist who is a mother of four children and Bruno's spouse; I am a human being who maintains a very specific frequency rate in the physical world. I am the expression of this frequency. All external restrictions sin against this essence. By choosing to express and feel my essence at zero point, regardless of the situation, I lighten up, I regenerate myself and feel my body vibrate more and more. Ascension is in fact precisely that: a body that vibrates at a new frequency and becomes multidimensional instead of being restricted by the earthly dimension.

Ascension and the Frequency Rate of Our Essence

By choosing to diffuse our essence throughout our interstitial void, we amplify our personal signature exponentially. To reach this point, we first have to learn to maintain our essential vibrational rate in daily life through the power of intention.

In learning to maintain the frequency rate of our essence with a certain consistency, we will also be able to command that our essence vibrate in the state of love. When these two tonalities meet, they form a third tonality, which is that of ascension. To ascend, the vibrational rate of self-love must be sustained by the frequency rate of our essence as this enables the density of matter to be modified.

Through continuity and stability, we will achieve a level of consciousness similar to that of the masters, whose frequency rate remains the same, regardless of where they find themselves. If Tibetan monks spend their lives preparing for their death, we should ask ourselves the following question: if we were to die today, at which frequency rate would we like to vibrate? Let us imagine that we leave our body in the vibration of self-love at zero point, and that this vibration is tuned on the frequency rate of our essence. We would no longer need to search for the tunnel of light when we die. The implosion in our essence would lead us to immediate ascension.

Nature is comprised of a variety of elements, each of which is endowed with a specific essence. Our essence is the energy point of our matter. It is constant and, among all the quantum possibilities that exist in Creation, it is our personal reference point. Our essence is directly linked to the Source through the Central Soul. We will try to amplify this frequency rate to discover how to maintain a vibrational velocity such that our body will change frequencies. The Central Soul will be able to help us with this, as we saw in the previous chapter. We will thus perform a masterful quantum leap as a result of the power of our intention. Once filled with our fundamental essence, what more do we need to become complete? Nothing! Nothing is more metaphysical or spiritual than to have our vital energy vibrate at maximum inside ourselves and to radiate it without influencing others.

Even if we hope that everyone will be invited to ascend, we need to be the ones to start the process. The critical mass for triggering such a reaction being about two percent, we must state the intention to succeed in maintaining a frequency wave that will set off a chain reaction. To be part of this first wave, our vibrational frequency rate needs to acquire its own identity so that each of our cells can vibrate at this essential frequency. When our body will have changed frequency rates, we can automatically share this vibrational information with others.

To ascend, we therefore have to choose to amplify our essential vibrational rate by maintaining that of self-love at zero point so as to achieve the vibrational velocity required to modify our body's frequency. When the body's frequency reaches a new modulation, which is specific to our essence and to self-love, it can experience a type of ascension that is different from Christ's, whose frequency rate reached a lighter tone before ascending into heaven. Quantum physics has proven that the nature of matter is vibrational. By commanding our body to vibrate at the frequency rate of ascension, we lead it to regenerate itself through the energy of self-love and essence. We will then feel our connection to the Central Soul and the Source. We will be free of time and space.

With the following exercise, you will learn the steps for making this a conscious process in your being.

Exercise 12
ASCENSION

You can use this exercise on yourself or as a therapeutic tool.

After answering the following questions and establishing your individual profile, you can choose from various tools such as prayer, meditation or DNA reprogramming to put the situation back at zero point.

If you choose DNA reprogramming, test all of the following questions to determine whether you obtain a yes or a no. Then, test to determine how many protocols you need to put these questions at zero point. (See DNA Demystified *and* DNA and the Quantum Choice.*)*

You can apply this exercise to a particular situation. If so, identify the situation.

TEST if the person knows that, to ascend:

1. her/his body must vibrate at a new frequency rate;
2. her/his body must vibrate at the frequency rate of self-love;
3. her/his body must vibrate at the frequency rate of her/his essence;
4. the tonality of her/his essence, when meeting that of self-love, will form a third vibrational tonality, a tri-tone, which is the frequency rate of ascension;
5. without being perfect, she/he can have the frequency rate of self-love and that of her/his essence vibrate by using the power of intention at zero point;
6. the frequency rate of self-love needs to be supported by the frequency rate of her/his essence so as to be able to change the density of matter;
7. she/he must see that matter is vibrational and that its frequency rate can fluctuate;
8. her/his body can become multidimensional while remaining in the earthly dimension;
9. she/he must state the intention of maintaining her/his essential frequency rate constant in daily life;

10. she/he must choose to amplify her/his essential frequency rate to discover how to reach the vibrational velocity that will enable her/his body to change frequencies;
11. she/he knows that, in this ascension process, her/his Central Soul and the Source are providing support;
12. she/he can carry out this process with:
  a. hope
  b. trust
  c. certainty
  d. conviction
  e. peaceful assertiveness
  f. faith in herself/himself
  g. faith in her/his essence
  h. assurance
  i. tranquility

The Influence of the Cosmos

As a result of planetary changes, variations in electromagnetic forces and solar storms, we are currently experiencing an important energy reversal. This **chaos** is helping to reshape our reality. It indicates new avenues of information, understanding and confidence required to accomplish the task that falls to us. This task consists of reacting according to the new paradigm, by virtue of which we can start using commands and intentions to reclaim our original frequency rate.

Since ascension is not a state but a frequency rate, it is possible that adjustments may be necessary before we can maintain a constant frequency. Like our planet, we are subjected to much vibrational turbulence. Indeed, the Earth is currently going through a crucial transitional stage in its evolution. It too is undergoing frequency modulations that affect us. The sun continuously ejects light particles, or photons, that can influence the awakening of awareness.

Our hypothalamus, like our pineal gland, is light-sensitive. These photons have a direct effect on the change in consciousness that we currently feel on the planet. The moment of conscious ascension is near and we are influenced by cosmic waves that trigger an awakening of awareness.

In its November 2004 issue, the magazine *Discover* published an article, on page 38, demonstrating that the spatial dimension is present on Earth and in our bodies. We are subjected to celestial influences without being aware of it. The Earth receives approximately 40,000 tons of meteoritic dust each year. At a subatomic level, the Earth is constantly bombarded with cosmic rays made up of particles from supernovas, among other things. It just so happens that the muons of supernovas can damage the molecules of our DNA and cause cancer.

In addition to its damaging electromagnetic effects, the sun emits ultraviolet rays that are responsible for the increase in skin cancer. In a less obvious way, the sun affects us through the neutrinos from its core. These electronic neutrinos bind themselves to the electrons in atoms. At night, they enter our bodies through our feet and come out through our heads after having gone through the Earth in 1/25th of a second. During the day, these neutrinos follow the same path in the opposite direction, from head to toes.

Constant changes to electrical charges affect the fields in which we live. Frequency modulations, which to a large extent are the result of photons, reach us in spiral waves. When they come into contact with the Earth, they submerge it all at once. These waves of particles leave as quickly as they come. In addition, at every moment, thousands of neutralinos (hypothetical cousins of neutrinos) are shooting through the most minute parts of our bodies. These neutralinos, which maintain the gravity of the galaxy, are apparently more numerous than all the stars combined. The neutrinos that surround the galaxies make up dark energy. They mingle with atoms and are very difficult to record. We are not aware of them having any adverse reactions. However, it is interesting to know that we are not isolated from the rest of the cosmos and that various rays and particles affect us without us being aware of it.

In an article on solar eruptions ejected from the sun, taken from the http://www.astrosurf.org/lombry/sysol-soleil-ha4.htm Website, it is mentioned that

> ...in the days following a **chromospheric eruption** [8], subatomic particles carried by the solar wind reach the Earth and enter the magnetosphere at the level of the poles, triggering brilliant auroras and radio blackouts in these regions. Such events, which do not occur with each eruption, disturb the

 8 Solar Flares

> Earth's magnetosphere, causing geomagnetic storms and resulting in overload at power plants, which can lead to the knock out of an entire electrical grid. Chromospheric eruptions are thus among the few astronomical phenomena that can directly disrupt the Earth's environment.[9]

Solar storms and winds, whose violence and frequency are increasing, reach us and interact with the atmosphere. One of the effects of solar winds is to increase the number of electrons in the ionosphere, which influences the speed of atoms. The relationship among electromagnetic particles is modified as a result and this change influences our awareness and our vibrational rate. The change in planetary awareness seems to be governed by solar and cosmic particles. Our metabolism is sensitive to light and to electromagnetic changes, and we are influenced by the full moon and by other astronomical and astrological events. In his book entitled *The Hidden Messages in Water,* which summarizes twelve years of research, the Japanese doctor Masaru Emoto demonstrated that water receives information and is sensitive to vibrational frequencies. Since our body is largely made up of water, we too are affected by the current planetary activation and change of frequency. We respond to cosmic forces through the process of ascension.

The Hypothalamus and Ascension

During my research on DNA reprogramming, I often came across data regarding the hypothalamus. It would appear that this organ is a universal translator of cosmic energies. I believe this is true since its biological function is directly related to the endocrine glands, and particularly the pituitary gland. It plays an important role in glandular secretions on the one hand, and the perceptual functions of the brain on the other. Along with the limbic system, it is associated with supernormal sensory functions and with the emotions. All information that enters the brain must first pass by the hypothalamus, which we can imagine as a blue pearl located in the middle of the brain.

The hypothalamus continually produces neuropeptides, which are the molecules of the emotions. These are small amino acid chains that vary over the course of the day according to the type of emotions being felt. The hypothalamus adapts to our style of thinking and regulates the quantity and type of peptides

9 Translated from the original.

released according to the particular vibration of each emotion. If we are angry, it will release a great number of peptides related to this emotion and will proportionally reduce the production of other sorts of peptides. Through the power of intention at zero point, we can re-educate our hypothalamus so that it secretes peptides that correspond to the vibration of our essence.

Just as the pineal gland, the hypothalamus contains a large number of photoreceptive cells that enable it to receive light information from the cosmos, which it sends to the nervous system and the brain. This is what makes it the best possible organ for telepathy and clairvoyance.

The hypothalamus governs the thyroid, the pituitary and the pineal glands. It reacts to our thoughts by continually producing peptides. In their turn, these peptides release proteins that influence our cells. The hypothalamus governs our repetitive behavior by transmitting emotional and neurological messages. The nervous system and the brain are directly linked to the hypothalamus. This organ plays a key role in our psychological state and our hormonal system. I strongly recommend that you watch the film *What the Bleep Do We Know,* which gives a very concrete picture of the role of the hypothalamus and the peptides.

In addition, the hypothalamus performs a relay function between the physical body and the Central Soul. It is an important actor in our process of transmutation toward conscious ascension. It is also the universal translator of cosmic forces. It filters universal influences and decides which ones can enter our bodies. It is a bridge between the body and the Central Soul and it is through the hypothalamus that the frequency rate of our essence can be amplified.

The following exercise will help us become aware of this little pearl nestled in the center of our brains.

Exercise 13
THE HYPOTHALAMUS

You can use this exercise on yourself or as a therapeutic tool.

After answering the following questions and establishing your individual profile, you can choose from various tools such as prayer, meditation or DNA reprogramming to put the situation back at zero point.

If you choose DNA reprogramming, test all of the following questions to determine whether you obtain a yes or a no. Then, test to determine how many protocols you need to put these questions at zero point. (See DNA Demystified *and* DNA and the Quantum Choice.*)*

You can apply this exercise to a particular situation. If so, identify the situation.

TEST if, to connect with her/his Central Soul so as to amplify the frequency rate of her/his essence in her/his interstitial void, the person knows that:

1. the connection must go through the hypothalamus;
2. the hypothalamus is a relay between the physical body and the Central Soul;
3. the hypothalamus can go through an energy transformation as a result of this reconnection;
4. this transformation can be expressed by a hormonal change triggered by the current planetary activation;
5. the hypothalamus will then be able to secrete new chemical components;
6. the hypothalamus will then be able to secrete new peptides that are in resonance with the frequency rate of her/his essence;
7. this transmutation of the hypothalamus will have a direct effect on her/his peptides and nervous system, which will be required to adapt to this biochemical change;

8. this biochemical transformation:

- **a.** will change her/him forever;
- **b.** will change her/his understanding of reality;
- **c.** will change her/his way of living;
- **d.** will place her/him on a quantum path that is in resonance with her/his essence ;
- **e.** will make her/his life easier and more comfortable;
- **f.** will activate her/his metabolism.

Activation of Ascension

One day, while I was in a ski-lift and was meditating on essence and the Central Soul, I had a very powerful revelation. Ski-lifts are a perfect place to think, especially when the snow is white and the sky such a pure blue! My feet were not touching the ground and I was between heaven and earth. That day I was entirely open to the sun and the cosmos when I felt the energy of my essence come directly from my Central Soul and revitalize my body. It was as though a ray of energy went through my hypothalamus and spilled over into my interstitial void and my organs. This connection transcended time and space and nourished my cells. I understood that if I wanted to connect myself to my Central Soul, which is the infinite reservoir of my essence, I could do so by going through my hypothalamus.

I knew that I could contact my hypothalamus by placing my tongue against my palate. This yoga technique is called khechari mudra. The word *khechari* has two parts: *kha*, the void, emptiness – sky; and chars, which means to move in, thus khechri means to inhabit the sky. Khechari Mudra is inverting the tongue backwards and touching the palate while fixing the gaze toward the eyebrow center. Having turned the tongue back, the three channels of *ida*, *pingala* and *sushumna* are opened. Yogic texts also proclaim that the tongue is then on the mouth of a well of nectar that seeps from the hypothalamus and is considered to be an elixir of immortality.

Following this experience, I developed a meditation that takes into account these elements and enables us to tune in to the frequency rate of our essence and of self-love, which forms the tonality of ascension. By repeatedly using this meditation, I came to understand and feel that I was a unique expression of Life in the great Whole. I realized that the conscious part of my identity was part of a vast reservoir of essence situated within the Central Soul, but oh! so easy to access!

Meditation on Ascension

This activation is done through visualization.

1. First of all, *command* to be at zero point, *even if* you do not know how. At the present time, our hypothalamus is going through great transformations caused by changes in planetary frequency.

2. Contact your hypothalamus by placing the tip of your tongue against your palate. The hypothalamus is like a little blue pearl shining in the center of your brain. The secretions of the hypothalamus taste like nectar.

3. Imagine that a corridor opens off the hypothalamus that connects you to your Central Soul.

4. Take the time to indicate to the Central Soul where you are in the universe. Imagine that you send a clear signal that bears the signature of your essence.

5. Imagine that your Central Soul is connected to the Source by another corridor of moving energy.

6. Visualize your interstitial void (±70% of our body).

7. Inhale and visualize your essence leaving the Source, heading toward the Central Soul and entering your body through your hypothalamus and your nervous system.

8. Let your essence amplify itself and vibrate within your interstitial void.

9. Open up to an intense feeling of love toward this essence that bears your personal signature.

10. Feel deep compassion toward yourself.

11. *Choose* to love yourself completely *even if* you do not know how.

12. Inhale and have the frequency rate of your essence vibrate at the same time as that of self-love.

13. Command that, from a combination of these two frequency rates, a third tonality–a tri-tone–superimpose itself on these two frequencies and begin to vibrate.

14. Command the complete amplification of this new vibration, which is that of ascension, within your interstitial void and especially in your heart and the center of your being.

15. Breathe and have this energy flow throughout your body, especially in your heart and the center of your being. Let it then spread to those around you: your family, your friends, your city, your state or province, your country and the Earth.

16. Finally, relax while knowing that you are supported in your process of ascension by your Central Soul and by the Source of Creation.

Conclusion

If we find happiness by tuning in to the frequency rate of our essence, we will start to have a concrete and kinesthetic experience of this vibration of happiness. We will know that everything is vibration and that emotions like happiness and pain are vibrational. We will feel that self-love and self-criticism each have their own frequency rate. From then on, instead of choosing the state of love at zero point, we will choose to **tune in** on the frequency rate of self-love at zero point.

We must expect that, unlike with our past experiences where we reached a limit in terms of our level of consciousness, our experience of our essence and of ascension can become continuous. The extraordinary new context resulting from this new experience will always bring us closer and closer to ourselves. By vibrating our essence in our bodies, we will feel greater and greater well-being. We will be regenerated by the frequency rate of our essence, which is entirely compatible with who we are. Of course this process does not unfold in a straight line; we can expect some ups and downs, but I am certain that it will lead us to much rejoicing. It is in commanding the amplification of our essence, despite our fear of not being able to do so and our powerlessness, that we will finally enjoy the prodigious power of our unique essence. We will ask that the frequency rate of our essence vibrate in our daily lives for eternity even if we are afraid that we will only lose it again and again.

Times have changed; we live in an era of acceleration. This amplification opens doors that enable us to access our essence. By using the magnetic charge of our separation from the Central Soul and the Source as our driving force, we can align ourselves with this expansion and choose, at zero point, happiness, the respect of our essential nature and that of others, the regeneration of our environment and our harmonious vibrational relationship with the rest of Creation. We can choose to use the magnetic power of conscious ascension to return to ourselves and build vibrational bridges that will encourage this unforgettable return within the network of quantum possibilities. This great cosmic return, whose hour has finally come and which has existed in probable futures since the beginning of Creation, depends upon a choice of frequency rate. We can easily make this choice through a magnetic intention at zero point. Let us use the full charge of the suffering caused by this long separation to command this return to our essence and vibrate at the same frequency rate as the Source. The age of separation is over; we are now able to discover our personal vibrational rate.

Unlike the classical spiritual model that places humans at the bottom of the scale and at the bottom of the tree of Life, this book on essence will end differently. I believe that we have come here to master the laws of density in order to transmit this information to the other dimensions. Of course, these angelical and etheric levels appear blessed and very light in comparison with our earthly dimension, which is heavy and strewn with contradictory events. But we must not forget that at these levels, the problem is precisely related to etheric worlds. We are the only ones who live in such a dense world. Consequently, we have become experts on density. We alone can voice an intention rooted in duality. By choosing conscious ascension through the tri-tone of our essence and self-love at zero point, we have opened a new avenue, which already existed among the quantum possibilities but which had never been materialized in the third dimension.

We cannot forget that we are experts on the earthly dimension. We are essential beings, who are very familiar with duality and the wavelength at which density vibrates, and we have to respect its needs. Through ascension, we will return along multidimensional wavelengths to teach the codes of duality and of four-dimensional (time and space) manifestations that are particular to this earthly dimension. We will need to have our humanity coexist with our divinity to transmit the laws of compassion and of zero point.

In conclusion, here is the comment I received from Jacinthe B. a few days after the workshop based on the content of this book: *"I have been taking workshops of all kinds for ten years, but this is the first time that I felt that all the elements were coming together… I hardly recognize myself anymore. I am calm; I no longer live on a volcano on the verge of eruption. In my opinion, this state comes from the combination of self-love, healthy boundaries, the recognition of my needs and, especially, the fact that I am vibrating in my ESSENCE. It is a state that is quite simply fantastic!*

And, like the song says, it's not over; the adventure has only just begun–this adventure which consists of living in self-love and at the maximum vibration of one's ESSENCE to produce the 3rd tonality, which is that of ASCENSION."

BIOGRAPHY

Professionally trained in naturopathy, Kishori Aird is a medical-intuitive practitioner. All medical-intuitive practitioners have their own methods of reading symptoms and re-establishing the vital current after identifying blockages. Ms. Aird is no different; she uses, among other methods, Reprogramming Kinesiology.

The path to medical intuition is taken more often nowadays than it was ten years ago. Ms. Aird has been on a long, spiritual and alternative journey. She lived in an ashram when she was 18. At 20, she took courses in nursing to become a midwife. At 25, after once again living in an ashram for two years, she started a family and moved to Ottawa where she worked at L'Arche de Jean Vanier (an international organization that works with people with developmental disabilities). In 1986, she returned to Montréal and started taking new courses in healing with crystals and hypnotherapy. Since the Harmonic Convergence in 1987, she has completed therapeutic training based on naturopathy and emotional therapies. She is trained in Tantra and is a Reiki Master.

In 1990, she was introduced to Reprogramming Kinesiology on the American West Coast and had the opportunity to work in a renowned chiropractic and naturopathic clinic. In 1993, after this training period in the field of naturopathy and kinesiology, she opened her own clinic and started working with clients as a professional in medical intuition.

In the summer of 1997, she began studies on DNA and the methods for reprogramming and reclaiming ownership of our DNA. She then went on to develop the reprogramming techniques that she teaches today.

Courses

Kishori Aird teaches, in both English and French, the reprogramming protocols found in *DNA Demystified* and *DNA and the Quantum Choice*. She also teaches a four-day workshop on the content of her book *Essence*. In these workshops, she introduces the protocols and teaches students how to make new connections and reprogram themselves using medical intuition.

For your information, you should know that there are no teachers currently authorized by Kishori Institute to teach the content of these books. To protect yourself, please consult us before registering for a training session on DNA Reprogramming or on the content of Essence. At this time, only Kishori Institute offers training courses.

For more information:

P.O. Box 252,
Magog (Quebec) JIX 3W8
CANADA
Tel.: (819) 868-1284
Fax.: (819) 868-9007
www.kishori.org
kishori@kishori.org

Suggested Reading

Berman, Bob. "The Cosmos Up Close. Outer space is not just out there—it is also on your windowsill and inside your body." *DISCOVER*, vol. 25 No. 11, November 2004, http://www.discover.com/issues/nov-04/departments/sky-lights/.

Emoto, Masaru. *The Hidden Messages in Water*, Beyond Words, Japan, 2001.

Chromospheric eruptions
http://www.astrosurf.org/lombry/sysol-soleil-ha4.htm

Dark matter
http://en.wikipedia.org/wiki/Dark_matter

Dark matter, cosmology, and large-scale structure of the universe
http://www.astro.queensu.ca/~dursi/dm-tutorial/dm1.html

Neutralinos
http://en.wikipedia.org/wiki/Neutralino
http://web.mit.edu/~redingtn/www/netadv/specr/6/node1.html

Neutrinos
History of neutrinos: http://wwwlapp.in2p3.fr/neutrinos/aneut.html

Science et Vie
"Le vide est plein d'énergie" June 2003, No. 1029.
"Sur la piste du code secret de l'ADN" December 2004, No. 1047.
"Dépasser Einstein" April 2005, No. 1051.
"Les câlins on un effet…génétique !" April 2005, No. 1051.

DNA Demystified

In a book entitled *DNA Demystified*, Kishori Aird reveals new information regarding DNA and how we can program it. More specifically, it is a matter of installing new programs or codes at places or addresses directly in our genes. In this book the author develops lists of instructions that she calls reprogramming protocols.

According to Kishori Aird, by functioning with 13 helixes rather than two, we are able to transcend a dualist understanding of reality and live at zero point–between shadow and light–in a circular perspective that brings with it compassion, understanding, abundance, humility and self-love.

DNA AND THE QUANTUM CHOICE

With *DNA and the Quantum Choice*, Kishori Aird invites us to enter the world of the infinitely small and to adopt a way of thinking that envisages the coexistence of multiple possibilities. In simple and precise terms, she explains concepts such as quantum understanding, the string theory, the theory of multiple worlds, the tunnel effect and non-linear time. In this way, she enables us to see that not only is the microcosm not static, but, to the contrary, it behaves differently depending on the observer. We learn that parallel worlds exist that vibrate at vibrational rates that are different from those of our world.

Kishori Aird proposes that we work with the protocols in this book and in *DNA Demystified* in order to be able to make new and enriching choices and make more judicious decisions. From now on we will enjoy the support of science and metaphysics. We have no reason not to use these new tools since our DNA is flexible and reprogrammable. By applying new scientific discoveries to human biology and the genetic code, we give meaning to life and matter.